MANAGING DRILLING OPERATIONS

ABOUT THE AUTHORS

Ken Fraser graduated from Newcastle-upon-Tyne Polytechnic with a Higher National Diploma in Mechanical Engineering in 1971. Following graduation, he joined Shell International's Drilling Department and spent thirteen years with Shell including four years on the brake, six years as Drilling Supervisor and two years performing office based duties. In 1984, he joined Houlder Marine Drilling and for two years worked as a Drilling Contractor Manager, operating semi-submersible rigs.

He has been Drilling Project Manager for single string ventures in Ireland, Portugal, Sweden and the U.K. and has managed drilling operations in Brunei, France, Germany, Holland, Italy, Malaysia, Norway, P.R.C., Spain, U.A.E., and the U.K.

He is currently Chief Executive Officer of North Sea Well Control Engineering Ltd (Norwell), the Aberdeen-based International Drilling Project Group. He lectures and consults internationally on Drilling Operations Management, Well Trouble Shooting and Well Control. He is an SPE member, author and technical editor.

Jim Peden is currently Shell Research Professor and Head of the Petroleum Engineering Department at Heriot-Watt University, Edinburgh. Formerly, he spent six years with Shell International in their Petroleum Engineering Department. He acts as a technical advisor to several oil companies and has worked in Brazil, France, Holland, India, Malaysia, Norway, P.R.C., U.K. and the U.S.A.

Andrew Kenworthy graduated from Glasgow University in 1987 with a BSc (Hons) Degree in Geology. He is currently a Drilling Engineer with Norwell in Aberdeen, responsible for Wellsite Engineering and Well Programming. He has worked in Ireland, Malaysia, Portugal, P.R.C. and the U.K.

MANAGING DRILLING OPERATIONS

by

Ken Fraser
(Norwell, Aberdeen)

with contributions from

Jim Peden
(Heriot-Watt University, Edinburgh)

and

Andrew Kenworthy
(Norwell, Aberdeen)

ELSEVIER APPLIED SCIENCE
LONDON and NEW YORK

ELSEVIER SCIENCE PUBLISHERS LTD
Crown House, Linton Road, Barking, Essex IG11 8JU, England

Sole Distributor in the USA and Canada
ELSEVIER SCIENCE PUBLISHING CO., INC.
655 Avenue of the Americas, New York, NY 10010, USA

WITH 22 ILLUSTRATIONS

British Library Cataloguing in Publication Data

Fraser, K. (Kenneth)
Managing drilling operations.
1. Fossil fuels. Extraction
I. Title II. Peden, Jim III. Kenworthy, Andrew
622.3381

ISBN 1-85166-630-3

Library of Congress CIP data applied for

Printed in Great Britain at the University Press, Cambridge

CONTENTS

For the Frasers; Jane, Liz, Madge, Al and Jack

Out of respect for Major __ de Coverley

ACKNOWLEDGEMENTS

I would like to thank Jim for writing Chapter 1 and Andy for Chapter 2, Catrina Flear our Administration Manager, who personally typed up the manuscript (many, many times over), the staff at Norwell and Jane, Liz and Al for all the support that they gave me during the preparation of this book.

I would also like to thank Per Arno of Corpro for his assistance in the preparation of the coring section of chapter 8.

Finally, I would like to thank the following who, over the years by their word and deed, have shown me how to manage drilling operations:

Fabrizzio D'Adda, Frank Allinson,
Greg Bourne, John Boor, Graham Buick, Pierre Bitzberger
Chen Yin Guan, Cheng Wai Ming, Chu Ping Ching, Peter Carson
Mike Donald, Lincoln Davies, Dave Deveney
Roger Easton
Mike Freeman
Richard Grey, Robbie Grant, Bill Guest, Peter Greaves
Pat Heneghan, Dave Harding, Geoff Hall, Brian Hatton, Keith Hewitt, Roy Hartley, Tore Hallberg
Francesco Illari
Dave Jarman
Randy Kubota
Li Kai Rong
Dave McKenzie, Marinus Maris, John McPherson, Neil Middleton, Donald McPhater,
Preston Moore, Megat Din
Raj Narayanan, Frans Notenboom, Dave Nims
Dave Parnell, Mike Pointing
Colin Rouse, Bouke Rienks, Derek Reynolds
Jaswant Singh, Syed Mohamed Bin Syed Othman, Mike Seymour, John Shute, Bill Stevens, Grant Schmit, Fokke Schroeder Snr., John Starling, Neil Simpson
Peter Thomson, Allan Tickle, Jimmy Turner, Ting See Lok, Mike Taylor
Willy Vermuelen, Joop Veldhoen
Gene Wilson, Willem Warmenhoven, Paul Waern, Bertis Wanningen
Xie Bang Qun

PREFACE

This book discusses all the technologies involved in managing drilling operations. Whilst looking at the obvious operational aspects of drilling oil and gas wells, it also tackles the less obvious but equally important fields of Contingency Planning, Contracts, Economics, Optimisation and Problem-solving.

A chapter is devoted to the creation of a Drilling Policy Document which can be used by the operating company as the back bone for its operations. Rig Selection and Acceptance is discussed in detail and finally Land and Offshore Operations are broken down into their component parts in a flow chart format.

To fit all this into a manageable sized text has meant making the assumption that readers are already familiar with drilling equipment and terminology. Furthermore, a multiplicity of units have been used in this book (reflecting current industry indecision on a standard). It is assumed that readers are conversant with these units or at least have access to conversion tables.

Chapter 1

THE ROLE OF DRILLING IN FIELD EVALUATION AND DEVELOPMENT

The evaluation and subsequent development of an oil and gas reservoir is a complex process which requires the integration of the skills and capabilities of a range of technical disciplines. The ultimate objective is to produce a plan for the development of the field and the subsequent implementation of that plan. As such, the process is one of iteration since, at the outset of the field development and evaluation, the amount of data is strictly limited and it is only as a result of activities to evaluate the field that information becomes more abundant and a clearer picture is obtained as to the reservoir and its production potential.

In general terms, the major objectives of field evaluation and development are as follows:

1. The identification, evaluation and confirmation of the following reservoir characteristics:
   (a) the nature of the hydrocarbons in place
   (a) the amount of hydrocarbons in place and the fraction which is recoverable
   (c) the productivity of the reservoir
2. The design, planning and installation of the wells within the field which will be necessary to allow the field to produce both economically and safely to satisfy the company objectives.

1. THE CHRONOLOGICAL BASIS OF FIELD DEVELOPMENT

The development of a field from initial exploration through evaluation and into subsequent development can vary substantially in length depending upon the size of the field, its complexity and the environment in which the field is likely to be developed. In some cases, particularly onshore, the cost of drilling exploration wells and conducting evaluation activities is relatively low and therefore it becomes

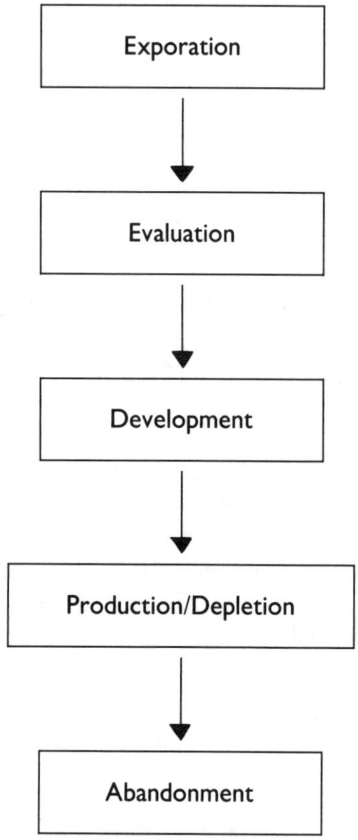

Figure 1 Chronological basis of field development

sometimes more easy to complete the evaluation process in a relatively short space of time, thus allowing the field to be developed. Conversely, onshore fields, whilst being easier to logistically support, are generally smaller in volumetric extent, therefore requiring less detailed engineering. In the offshore environment, the difficult logistics and the need for more advanced technology may make developments uneconomic unless they are of a substantial size.

The various phases for the development of a field are shown in Figure 1, and it can be seen that they pass from exploration through evaluation and development, to subsequent production and depletion, and finally the abandonment of individual wells and the field. Each of these phases will be discussed in turn below.

1.1 Exploration

The major objective of the exploration phase is to identify the prospect in structural terms. In this context, it will be necessary to produce a physical map of the subsurface structure. The objectives are to identify the presence of a suitable structure subsurface which will be likely to act as a trap for the hydrocarbons in moving from the source to the reservoir rock. It is important to realise that the structure identification is on a macro scale since data available to make such prognosis is relatively limited. In addition, whilst it is possible to identify suitable structures which may act as a trap to retain oil and gas subsurface, there is no attempt at this stage, nor is it possible, to confirm the presence of hydrocarbons within that structure. However, indications may be forthcoming from geological prospects in adjacent blocks or in adjacent areas where more detailed information is available.

The work of exploration is largely conducted by geologists and geophysicists. In general terms, the identification of the structure is accomplished by some of the following techniques:

1. **Previous geological/geophysical information** In general terms, more reservoirs are encountered in sedimentary basins whereby general depositional trends can normally be used to infer the presence of other reservoir sites e.g., the North Sea. This technique of comparing data from adjacent prospects for which more data is already available can be an extremely useful indicator as to the potential for other reservoir systems to exist.
2. **Seismic** The process of seismic surveys involves the creation of an acoustic sound wave which is directed at the subsurface. The movement of the seismic wave subsurface depends upon the density of the strata and also changes in the physical structure of the layer's subsurface. The acoustic waves will therefore be reflected by changes in lithology, the reflected wave is collected and the signal response analysed to produce a sectional map of the strata subsurface. During the past decade, significant advances have occurred in the area of seismic. Three-dimensional seismic modelling has substantially contributed to exploration capabilities.
3. **Gravimetric surveys** In a gravimetric survey, an attempt is made to measure the change in the earth's gravitational field. Such changes in the gravitational field will be influenced by the minerology subsurface and in particular the presence of salt plugs. Salt plugs, which are plastic in nature, will generate a series of traps within the subsurface structure and these can act as potential sites for the accumulation of reservoir fluids.
4. **Satellite surveys** Growing use is being made of satellites with infra red imaging techniques to detect potential subsurface deposits of hydrocarbons and other minerals. This technique will find economical application in many areas of the world and is likely to be more widely applied by the oil industry in the future.

The exploration phase will identify, hopefully, the presence of a suitable structure which may act as a trap preventing the upwards migration of potential hydrocarbons. Importantly, the exploration phase will yield a geological/geophysical map of the structure which will indicate the approximate size and perhaps the extent of faulting within the potential reservoir unit. There is little opportunity, at this stage, to make any further suppositions as to the presence or not of hydrocarbons. The only means by which the presence of hydrocarbons can be evaluated is to actually penetrate the structure physically and measure and sample the reservoir accordingly.

1.2 Evaluation

The process of prospect evaluation is both expensive and will require considerable technical resources to effectively evaluate interesting reservoir prospects. The information required from an evaluation process consists of the following:

1. Identification of the depositional sequence within the reservoir and information relating to the thickness and minerology of the various sediment layers.
2. Information relating to the pore space in terms of the porosity and the fluid saturations within the pore space.
3. Information relating to the permeability or production capacity of the reservoir unit.

The above list is not exhaustive but is intended to indicate the production capacity of the reservoir unit.

It therefore becomes clear that the evaluation of the reservoir will require wells to be drilled to penetrate the structure at several different areal locations. In this way, information will be provided to assist in more detailed geological mapping of the structure. The evaluation process will require inputs from the Exploration Department as well as Drilling and Petroleum Engineering. The co-ordination of information collection and data acquisition is normally the responsibility of the Petroleum Engineering Department.

Drilling exploration wells will generally be conducted with specific objectives in terms of data acquisition and these will be defined at an early stage. Further, the cost of exploration wells, in many cases, must be written off against the value of the data which is acquired for reservoir evaluation.

In the drilling programme for an exploration well, a number of evaluation activities will be built into the programme. These evaluation activities will take place through the reservoir while it is being drilled. Information can be obtained by the following methods:

1. **Coring**, whereby a cylindrical section of the vertical sequence of the layers in the reservoir is cut and retrieved for surface evaluation. In this technique, the principal objectives are to obtain a large sample of the reservoir rock, with detailed information on the sedimentary sequence in which the rock system exists.
2. **Logging** with wireline will yield considerable information in relation to the borehole, the near wellbore reservoir area and the fluid content in that region. Various logging systems are available, including:
   (a) acoustic logs which can be used to evaluate the porosity of the reservoir rock system.
   (b) nuclear logs which can be used to identify the porosity and the type of fluids within the pore space. Nuclear logs can also be used for a variety of other reasons, including the determination of sand stability etc.
   (c) resistivity logs which will yield information on the ability of the rock pore space and insitu fluids to conduct an electrical current and will therefore yield information in relation to the fluid saturation within the pore space.

(d) resistivity induction tools which will provide information on the ability of the drilling fluid to flush through the pore space. This information will yield valuable insight into the ability of the reservoir fluids to be flushed from the pore space and this will have a direct bearing on the potential recovery of hydrocarbons from the reservoir pore space.

3. **Well testing** will largely provide information in two areas:
   (a) the presence of mobile hydrocarbons by inducing the reservoir to produce fluid towards the wellbore.
   (b) information relating to the production capacity of the individual well and in particular, the relationship between pressure across the reservoir and the production rate.

Well tests can be conducted by using wireline tools, or using either a drill stem or a production string to flow test the well. Whilst testing is an expensive process, it can be used, if designed correctly, to generate information which is representative of a wide section of the reservoir unit.

1.3 Development

Once the information has been gained from the exploration and evaluation phases and the reservoir has been modelled to identify the amount of hydrocarbons which can be recovered, the reservoir development will be designed and the economics evaluated. At this stage, it is important to realise that limited information may be available but if the economics permit, the development will proceed and a number of development wells will be drilled and completed. During the drilling of the development wells, the wells themselves will be evaluated to yield further information in relation to the reservoir, rock and fluid properties at the specific location of the individual wells. Therefore, in the development drilling phase, a considerable amount of evaluation will be involved and this will be used to improve the reservoir model and to monitor the development and subsequent completion of the reservoir.

The development process can therefore be viewed as being not only essential but also yielding further information on the reservoir which will allow the development plan to be modified dynamically as the development proceeds. Of particular interest in the development phase will be the following:

1. The drilling and completion of individual wells.
2. The means by which the reservoir is completed across the production interval.
3. Platform requirements for wellhead flowlines and fluid separation.
4. Fluid processing and export systems.

A considerable amount of work has to be expended in the development phase to correctly evaluate the reservoir development strategy and of particular interest here is the assessment as to how the reservoir will respond dynamically and over an extended period, to the process of production. It may be necessary to consider the possibility of supporting reservoir production capacity by using fluid re-injection for pressure maintenance. Alternatively, it may be necessary at some stage in the life of the reservoir to institute a process of artificial lift whereby the production of fluids from the reservoir will be assisted.

An important feature of the development phase is that it may last over a period of several years and in fact will overlap with the subsequent production phase. Therefore considerable technical effort has to be expended on continuously monitoring reservoir production and well performance.

1.4 Production

The production phase of a development will be intended to allow oil and gas production to proceed as follows:

1. At the maximum rate and for the maximum period possible.
2. The production phase must at all times ensure maximum safety in view of personnel, capital costs and the environment.
3. There must be a continuous assessment of the production process to maximise the efficiency with which it is conducted and to ensure that minimum production costs ensue.

The production phase is therefore one of considerable importance for the overall economics of the development. It must therefore involve the application of technical skills, not only to maintain production, but to improve the production process and efficiency.

1.5 Abandonment

The performance of individual wells will be continually monitored and periodically assessed to identify ways in which their performance could be improved and their production made more economic. At some stage in the production of the well, a point will be reached whereby the well can no longer produce oil or gas economically i.e., the cost of the well and its production will exceed the revenue arising from fluid production.

2. OPERATING COMPANY ORGANISATIONAL STRUCTURE

The means by which exploration, drilling, petroleum engineering and production interface within an oil company varies between companies. In some cases, these specialisms form distinct departments within the organisation, whereas in others, the structure evolves from a limited number of departments and therefore would involve some combination of specialisms, such as exploration and petroleum engineering or well services and production.

A typical structure is indicated in Figure 2 which shows the existence of five separate departments within the structure. In this particular example, these departments are

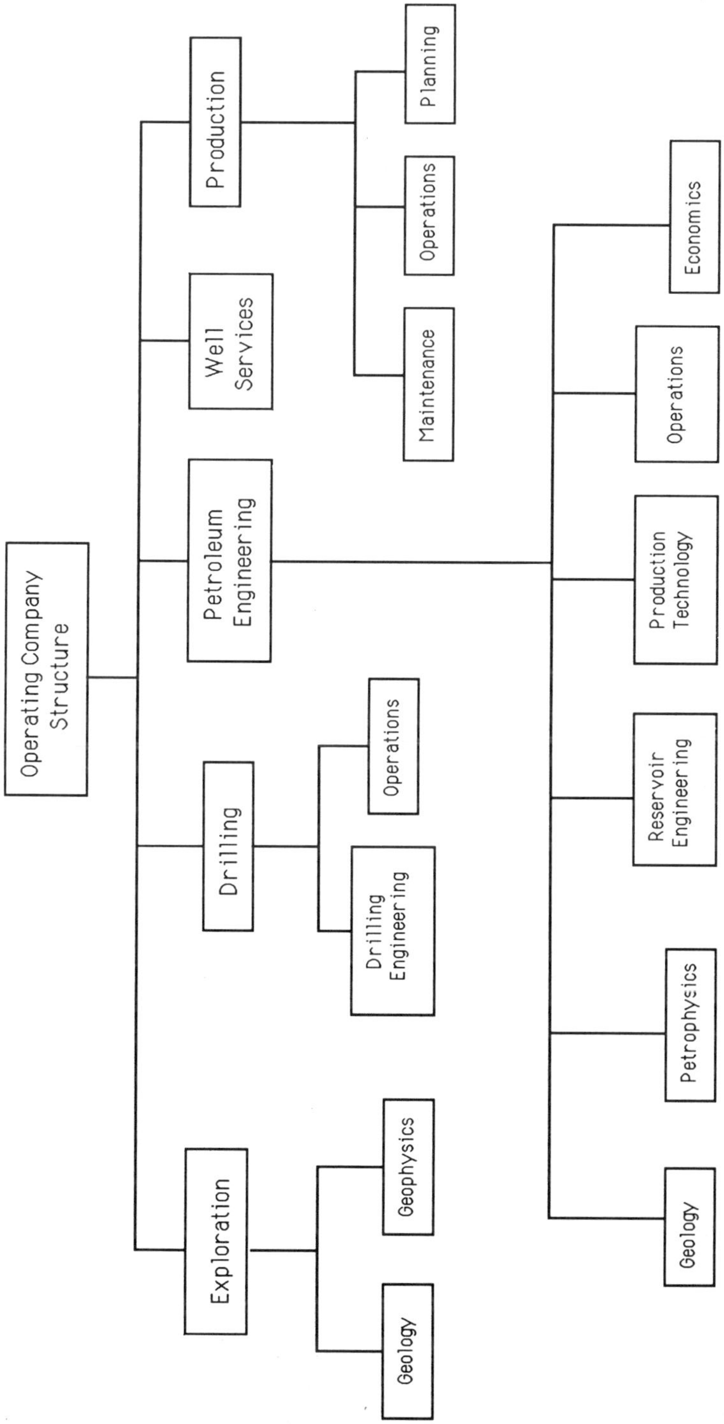

Figure 2 Structure of a typical operating company

Exploration which includes Geology/geophysics; Drilling which includes Drilling Engineering and Drilling Operations; Petroleum Engineering; Well Services; and Production which comprises Maintenance, Operations and Planning. It can be seen that the range of disciplines involved in petroleum engineering is quite extensive and in many situations, this broad range of capabilities is used to co-ordinate across the time span of the exploration, development and production phases.

2.1 Exploration

The Exploration Department will be responsible for identifying structures for consideration for development and providing a substructure map of the prospect. The responsibility of exploration would be to further update, refine and modify the substructure map and reservoir modelling in accordance with the increased amount of data which becomes available during the development programme. The Exploration Department will further be required to provide guidance on the selection of final well locations in the development plan in conjunction with the Reservoir Engineers, within Petroleum Engineering, who will be assessing the recovery of oil or gas from the structure as a function of the final well locations.

2.2 Drilling

The Drilling Department is responsible for the safe and efficient drilling of the well to defined targets and locations identified by Exploration and Petroleum Engineering. They are further charged with the responsibility of ensuring that all evaluation work is conducted safely and in accordance with the requirements of the other departments. In this context, there will generally be two specific functions within drilling, namely, Operations, which are responsible for the day-to-day supervision and planning of individual wells, and Drilling Engineering, which will be responsible for the adaptation and development of new or improved technology for inclusion in the drilling programmes.

2.3 Petroleum engineering

Petroleum Engineering is a broad based discipline which has a prolonged input to reservoir evaluation and development.

2.3.1 Petroleum geology

Normally, there will be geological specialists within the department, who will work closely with the Petrophysicists and Reservoir Engineers to ensure that locations of individual wells and the evaluation process is carried out efficiently and yields the required information to improve the reservoir model developed by the company.

2.3.2 Petrophysics

A Petrophysicist is responsible for recommending the wireline logs which will be run into individual wellbores and for the analysis of those logs to yield information relation to the reservoir structure and fluid composition. This function is therefore crucial to ensuring that the exploration and development wells yield the required information to provide detail within the geological structure model.

2.3.3 Reservoir engineering

Reservoir engineering is a broad discipline and as such Reservoir Engineers will be responsible for the following areas of technology:

1. The properties and performance of reservoir fluids.
2. The response of the reservoir rock to the production process.
3. Assessment of the response of a reservoir to the production or depletion process.
4. To identify and recommend the means by which oil recovery can be enhanced or improved e.g., pressure maintenance or by the use of enhanced oil recovery.

In general terms, the Reservoir Engineer is charged with the responsibility of ensuring that the reservoir can be exploited as effectively as possible and that the reservoir energy available within the fluid is fully utilised to maximise the potential recovery from the reservoir.

2.3.4 Production technology

The Production Technologist or Engineer, is responsible for the wellbore and the completion equipment installed within it and also with the consequence of production in terms of the reservoir fluids e.g., the tendency for scale, wax or asphaltene deposition. In the cycle of reservoir evaluation and development, Production Technologists with be heavily involved in the design and selection of equipment which will be installed inside the wellbore and which will be required to withstand operating conditions and the fluids. In the longer term development of the reservoir, the Production Technologist will be charged with maintaining the wells at their peak operating efficiency and ensuring that maximum recovery is achieved. This may necessitate the implementation of workovers to correct mechanical or reservoir problems which may arise as a result of continued production.

2.3.5 Operations

The Operations Section within Petroleum Engineering provides the necessary link between operational groups within Drilling, who will be responsible for the drilling of the exploration and development wells and the evaluation and technical specialists within Petroleum Engineering for whom the well is being drilled to yield the necessary information for the reservoir modelling. The Operations Section, therefore, requires a detailed understanding of the role of drilling and also of the various disciplines within Petroleum Engineering to ensure that they can provide the effective co-ordination necessary.

2.3.6 Economics

The role of economics is fundamental to both the evaluation, development and abandonment of reservoirs and wells. It is seen as being the means by which technical information can be transmitted into management terms to allow decisions to be made regarding future investment or abandonment of projects.

2.4 Well services

The design of most development well completions utilise, to a varying extent, slick wireline activities. The role of Well Services is to specify and prepare completion equipment for installation inside the wellbore and then to periodically conduct repair work within the wellbore to replace malfunctioning components.

2.5 Production

The Production Department is responsible for the ongoing and continuous production of fluids from the reservoir. Their responsibility is therefore to monitor and control production in such a way as to maxmise the recovering of reserves from the reservoir. The planning of production rates and production plateaus is frequently based upon reservoir models generated by Reservoir Engineering within the Petroleum Engineering and will be implemented by the Production Department. Since the Production Department is responsible for the development wells once they are in production, it is their responsibility to ensure the wells are maintained in peak operating capacity and as such they will be responsible for co-ordinating all maintenance work required within the platform and also around the individual wells.

Chapter 2

AN INTRODUCTION TO GEOLOGY FOR DRILLING TECHNOLOGISTS

1. THE RELATIONSHIP BETWEEN GEOLOGY AND DRILLING TECHNOLOGY

Geology is the study of the earth as a whole, its origin, structure, composition and history (including the development of life) and the nature of the process which gave rise to its present state. Geology encompasses the processes which form the medium through which the engineer drills. To understand the problems of drilling, it is necessary to understand the nature and formation of the material drilled.

2. GEOLOGICAL TIME

The idea of geological time is perhaps the first concept to master. In every drilling programme, there is at least the framework of a stratigraphical column. To the non-geologist, often this does not mean much, but in fact it is a fairly simple concept. The earth has existed for 4600 million years, which spans the entire geological time as we know it, however, hydrocarbons are rarely found on any rocks older than Cambrian, which is 500–600 million years old.

This time span 0–4600 million years is divided into sections which are given names. The largest of these sections being Earatherm, e.g., the Cenozoic or Mesozoic. Earatherms are subdivided into systems with names such as Permian, Jurassic, Cretaceous etc. These terms are the most common and well known periods used in geology. Again these are subdivided into series i.e., Upper and Lower Jurassic and stages within series e.g., Kimmerage and Oxfordian.

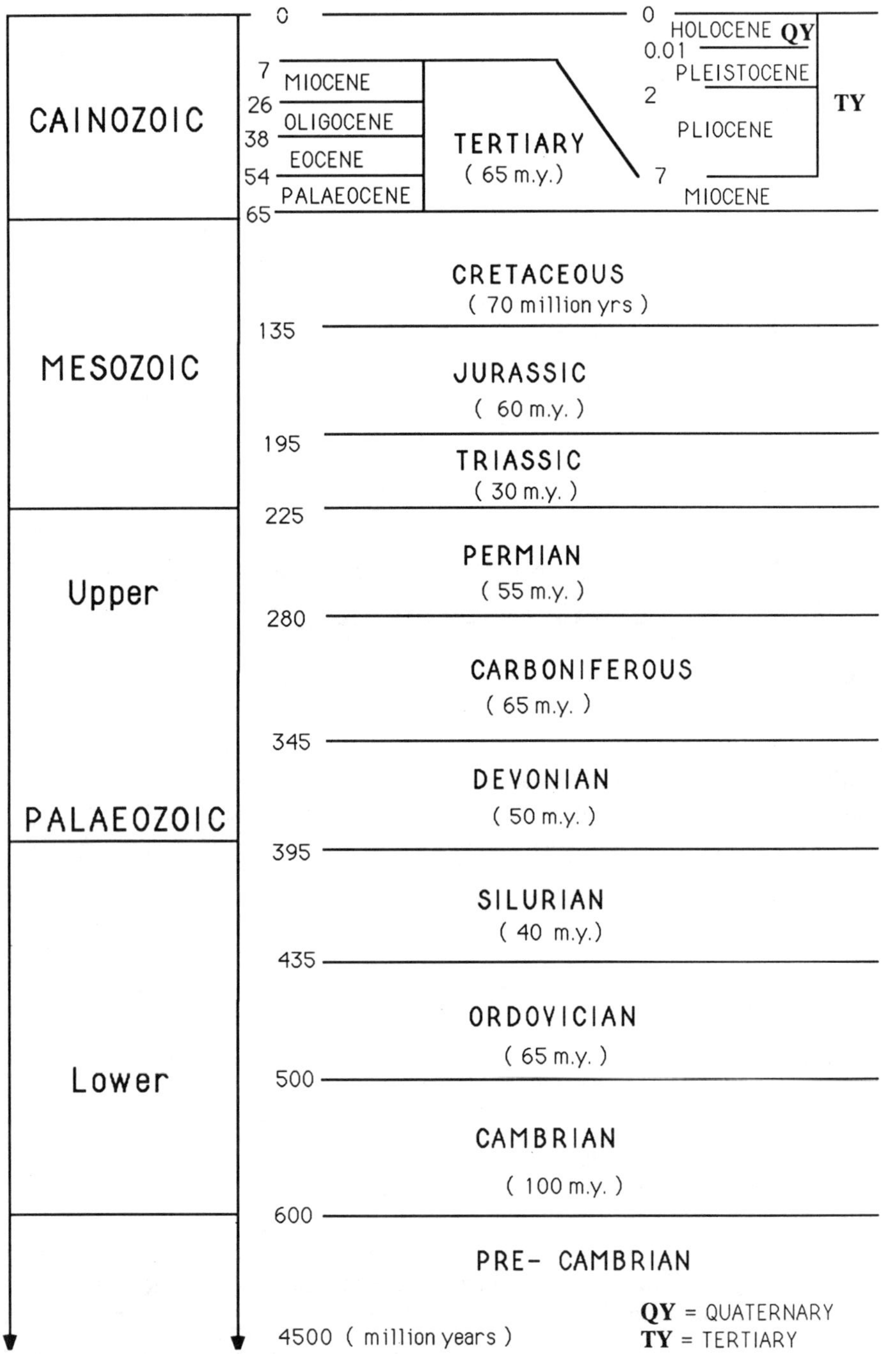

Figure 1 Geological time chart

3. ROCK TYPES

Rocks can be divided into and described in three main groups, igneous, sedimentary and metamorphic:

3.1 Igneous rocks

These can also be described as primary rocks. They are formed from molten rock (magma) and crystallise from this melt as combinations of minerals. These combinations of minerals are related to the initial melts chemical composition and the crystallisation pressure temperature regime. Common igneous rocks are granite and basalt. The names given to igneous rocks relate to their (crystallised) constituent minerals, crystallised mineral grain size and texture. Igneous rocks are often intruded or injected as a melt into existing rocks along faults, joints etc., often following the path of least resistance. After injection, they solidify to form crystalline rock. A confusion that arises when dealing with igneous rocks is the difference between magmatic and volcanic activity. Magmatic relates to molten rock below the earth's surface, whereas volcanic relates to molten rock after extrusion.

3.2 Sedimentary rocks

Sedimentary rocks can be described as secondary rocks and there are two basic sources:

1. Deposits which are made up from the remnants of pre-existing rocks. Sedimentary rocks of this type are made of the remnants of pre-existing rocks through erosion of rocks by chemical and mechanical systems e.g., freeze-thaw, river complexes, hydraulic fracturing, sand blasting, chemical solution etc. This breakdown means that components of the existing rocks are broken from an existing body of rock and transported by various methods i.e. gravity, water flow, wind etc. These fragments of rock are carried to another site, where they are deposited. As time passes, they build up into layers and become buried forming new rock which is sedimentary. Sources of sedimentary rocks can be igneous, metamorphic and previously existing sedimentary rocks.
2. Chemical and biological precipitates. Examples of chemical precipitates could be salt or gypsum which are formed as evaporates in hot climates. This usually takes place in shallow lagoonal water where the influx of water is less than evaporation. Other chemical precipitates include types of limestones which are formed by changing the physical condition of water saturated with $CaCo_3$.

   Biological precipitates may be limestone reefs which are particularly important as reservoirs. Coral is an animal which secretes $CaCo_3$ to give it its structure. When coral dies, new coral builds on top, thus building up a reef. Coal is also formed from organic matter which is deposited in deltaic conditions in hot climates and buried quickly before it has a chance to degrade.

3.3 Metamorphic rocks

Metamorphic rocks can be formed from sedimentary, igneous or previously existing metamorphic rocks. Metamorphism is the change from one state to another. Rocks that have been metamorphosed have undergone at least one of two processes. These are change in temperature and change in pressure. The pre-existing rock's chemistry may change due to partial melting and either loss or gain of chemical components. The gain of components usually involves the addition of water, i.e. mineral hydration, whereas loss of components is usually loss of water, i.e. dehydration. Other types of metamorphic changes occur in closed systems and are due to pressure and temperature with different structured minerals forming from the same chemical components. Almost any rock that has been changed by temperature and pressure could be described as metamorphic.

There are two types of metamorphism: progressive and retrogressive. Progressive metamorphism involves an increase in temperature and pressure. Dehydration occurs as minerals become more dense and water is lost. Regressive metamorphism involves a decrease in temperature and pressure with the addition of water. Non hydrous, or partially hydrous minerals break down to form more hydrous minerals. The most hydrous minerals are clays which can adsorb large quantities of water.

In many metamorphic rocks formed from sediments, remnants of the previously existing sedimentary structures may exist dependent on the temperature and pressure conditions of metamorphism. Pure quartz sandstone, when undergoing progressive metamorphism, does not tend to change composition because of the stable nature of quartz (SiO_2). Effects of metamorphism can be seen in the internal structure of the rock which forms a bonded texture. Pressure solution dissolves silica and then with changing pressure/temperature conditions recrystallises it, bonding the grains together. This is how hard quartzites can be formed.

3.4 Sedimentary rock type and structure

Due to the fact that most hydrocarbons produced in the world today are reservoired and generated in sedimentary rocks, it is perhaps a good idea to study them in more detail.

Most sedimentary rocks are stratified or bedded i.e., occur in laid down layers. Each layer is separated from one another by a bedding plane. The attitude of this layer i.e., its dip and strike, is fundamental to interpretation of structures that may be present in rocks.

Each type of sedimentary rock is formed in a set sedimentary environment, which is related to the physical and chemical conditions under which it was deposited or formed. Many different types of environment can exist at the same time in different topographic and climactic zones. These zones are characterised by a different set or combination of rock types:

1. Deserts form dune sandstone e.g., Permian Rotliegendes, formation of the southern North Sea.
2. Shallow warm seas form carbonates e.g., many of the Middle East large reservoirs.
3. Deltas form large volumes of land derived sediments e.g. Niger and Mississippi deltas.

Environments change with time. Continental conditions may be replaced by marine conditions and vice versa due to fluctuating sea levels. Fossils can give indication of the enviroment in which a rock was deposited as well as rock type. Differing fossil communities also show subtle variations in environmental conditions. Initial environmental conditions can influence whether a formation has the potential to source or reservoir commercial hydrocarbon deposits.

The common sedimentary rocks in relation to the oilfield can be classified as follows:

sandstones	reservoirs
carbonates	reservoirs
shales	source and cap rocks
evaporates	caprocks
coal	potential source of gas

3.4.1 Sandstones

These are formed from rock fragments measuring 0.06 – 2 mm. Their most common constituent is quartz SiO_2 followed by silicates of Al, K, Na, Ca etc. Sandstones are generated in a wide variety of environments:

fluvial environments (rivers, streams etc.)
delta fronts or channels
coastal plains, barrier island, tidal channels
desert and coastal aeolian plains
shallow and deep marine environments

About half of the world's total recoverable reserves of oil and gas occur in sandstone reservoirs.

3.4.2 Carbonates

There are two main types of carbonate ($CaCO_3$ and $CaMg\ (CO_3)_2$ limestone dolomite). Although clastic limestones do occur (derived from the erosion of pre-existing carbonates) most limestones are of chemical or biochemical origin:

1. As a bi-product of the life process of animals or plants.
2. Direct chemical precipitation from sea water.

Precipitation of $CaCO_3$ occurs in warm, clear, shallow water away from silicate detrial deposition. Limestone is deposited under limiting temperature and depth conditions. Coral needs sunlight to grow and is therefore deposited only in shallow conditions. Generally, a water temperature in excess of 25°C is needed for limestone deposition as $CaCO_3$ is much more soluble in colder water.

Dolomite formation is a controversial issue and the debate is centred on whether primary (direct chemical precipitation of $CaCO_3$) or secondary precipitation occurs.

Is dolomite deposited directly, or does it occur due to mineral dissolution and replacement by percolating fluid solution? Evidence suggests that both processes occur.

Note: Chalk is a very fine grained pure limestone found in the Upper Cretaceous of Western Europe however the term is occasionally used in other areas of the world for similar fine grained limestones.

Limestone and dolomite reservoirs contain approximately 50 per cent of the world's total recoverable reserves of oil.

3.4.3 Shales

A shale is a fine grained detrital rock composed of silt and clay particles less than 1/16 mm in diameter. The most important components of shales are fine crystalline silicates of Al, Na, K and Ca with quartz, calcite and dolomite making up most of the remainder.

Organic rich shales deposited under anoxic conditions can act as source rocks under favourable conditions. Due to their very low permeability, and semi-plastic nature shales also function as cap rocks or seals to oil and gas accumulations.

3.4.4 Evaporites

Evaporites are chemical precipitations from concentrated solution or brine. Their formation requires greater evaporation than influx of water, which tends to only occur in arid conditions. The most common evaporite types are as follows:

anhydrite	$CaSO_4$
gypsum	$CaSO.2H_2O$
rock salt	NaCl

Evaporites are the most efficient cap rocks because of their impermeability and plastic nature.

3.4.5 Coal

Coal is formed from dense forest close to the coastline, building up a layer of plant material faster than decomposition can occur. This is followed by a change in sea level, which brings an influx of salt water killing off the root zone. Continued subsidence allows sediments to bury the un-decomposed plant material, sealing it from the atmosphere and gradually compacting it. If subsidence stops, erosion takes place followed by the growth of a new layer of plant material. Subsidence is then reactivated with an influx of sea water causing the process to repeat. This cycle is referred to a cyclotherm and it is the main deposition mechanism for coal deposits.

The quality of coal, (its calorific value) increases with maturity, that is, depth of burial, heat and compression. Much of the southern North Sea gas is produced from coals of Carboniferous age and reservoired in Permian sandstone.

4. STRUCTURAL GEOLOGY

Structural geology is the result of tectonic stress. Structures vary in size from regional (hundreds of kilometers) to micro fractures (millimetres). Each type of structure is significant in structural geology as the same basic structures occur throughout a tectonic area, but on different scales. Almost all structural interpretation is based on combinations of a few basic concepts.

Faulting, folding, fracturing, thrusting, are all terms used in structural geology. These terms are used to describe movements of bodies of rock after or during deposition. These are the basic processes that form many of the world's hydrocarbon reservoirs.

4.1 Faulting, folding and fracturing

In general, deformation is due to changes in tectonic stress, which can manifest themselves in a number of ways. The modification of a rock can be represented by the Bingham plastic model in terms of stress and strain, (Figure 2).

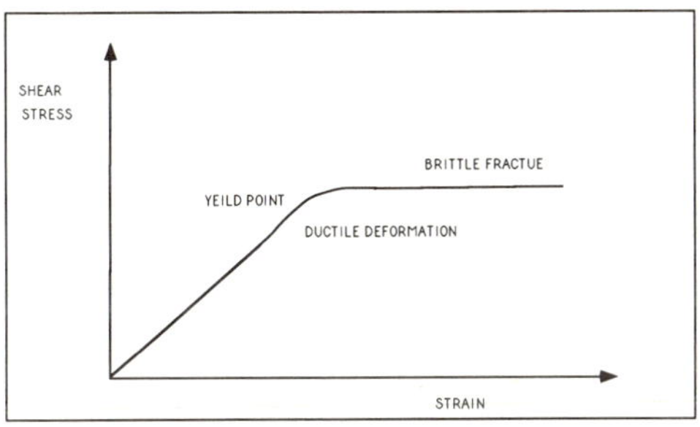

Figure 2 Bingham plastic model

Depending on individual properties of rocks and the external forces upon them, folding (ductile deformation) or faulting (brittle fracture) will result. The yield point is the point at which elasticity of the rock is overcome and permanent deformation results. The yield point varies for different rock types and also different pressure temperature conditions.

4.1.1 Faulting

Stress on a body can be categorised into three components for faulting: S1>S2>S3, (Figure 3).

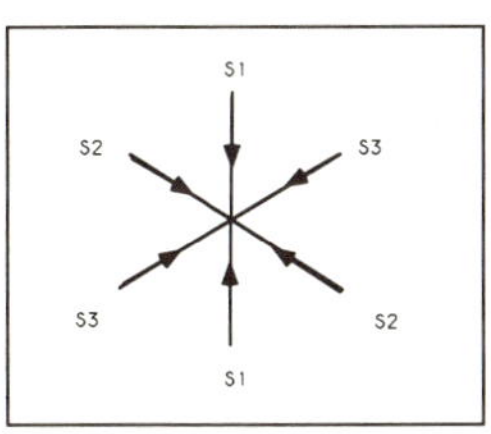

Figure 3 Stresses on a body

For a normal fault (extension fault), the stress regime is shown in **Figure 4 below.**

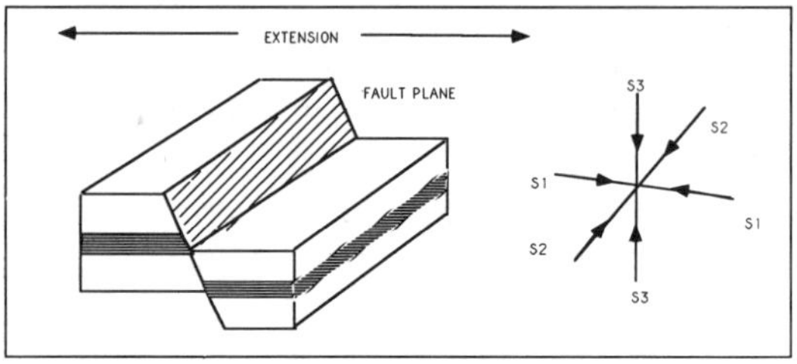

Figure 4 Normal extension fault

The greater difference between components S1 and S3, the more likely faulting will occur. This property can be related to the yield point.

The stress regime for a reverse fault (compression) is shown in Figure 5.

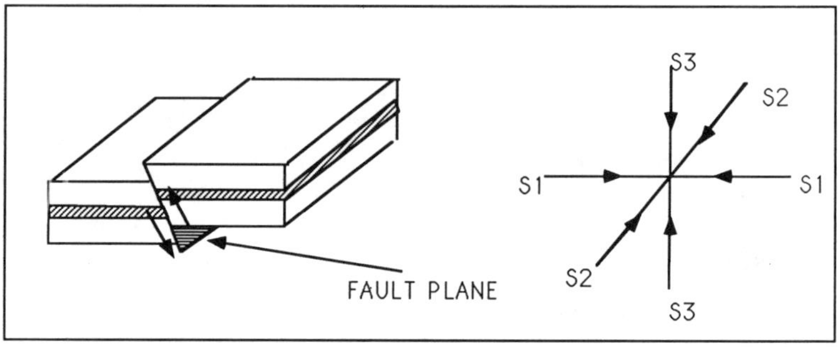

Figure 5 Reverse compression fault

The stress regime for a strike slip fault is shown in Figure 6.

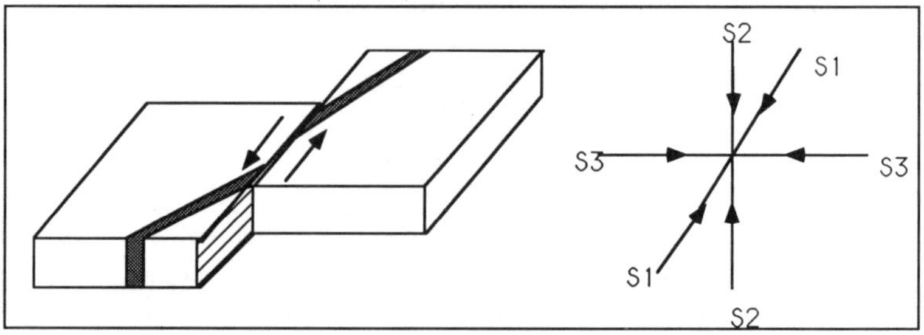

Figure 6 Strike slip fault

The angle or inclination of the fault is related to rock properties, the direction of the stress field and the relative strength of each component.

4.1.2 Folding

Folding occurs due to compression of a sequence as shown in Figure 7.

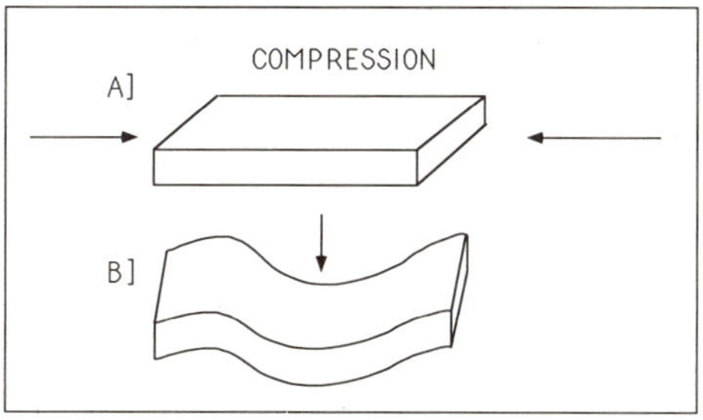

Figure 7

The amplitude and wavelength of the fold is related to the competence and the thickness of the sequence being folded (Figure 8).

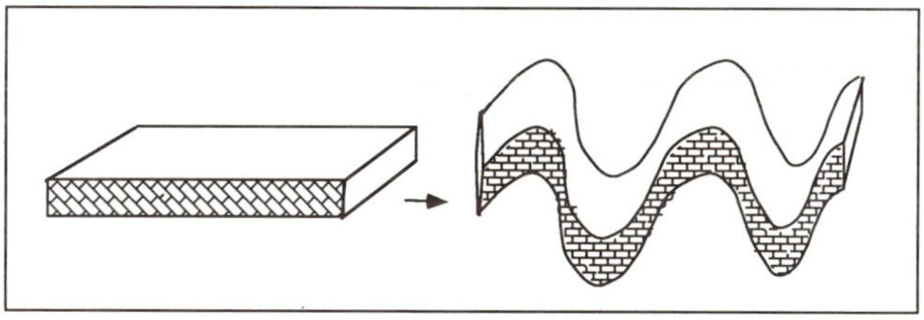

Figure 8

Folds are produced by crumpling, buckling or arching of strata. An anticline is an arch in which two circles, usually limbs or flanks dip away from each other, (Figure 9). A syncline is a fold in which the limbs dip towards each other, (Figure 10). A monocline is a steplike fold in which horizontal beds locally become dipped and then flatten out, (Figure 11).

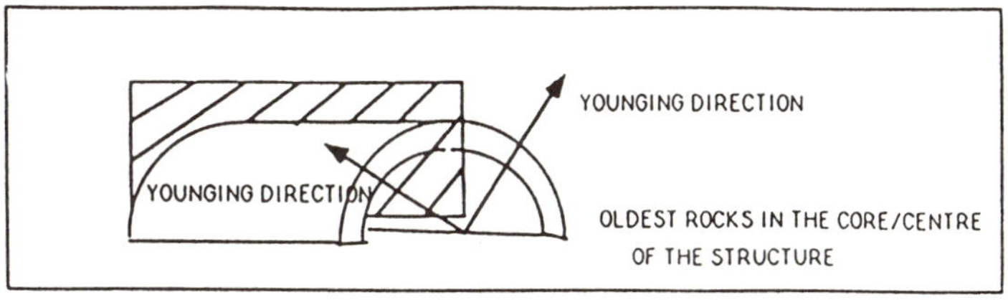

Figure 9

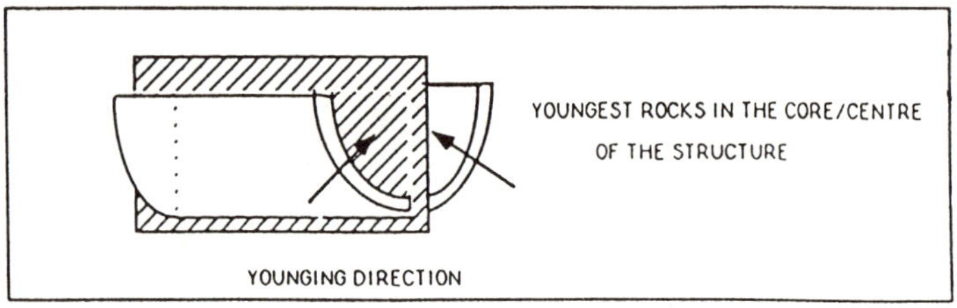

Figure 10

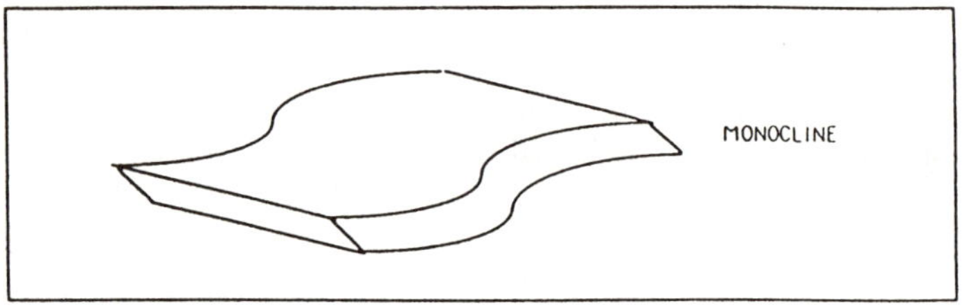

Figure 11

4.1.3 Joints and fractures

Joints and fractures have little or no displacement and are usually on a small scale when compared to faulting and folding. They often occur in homogeneous rocks and relieve stress throughout a body rather than manifest the forces into faults. An example of joints and fractures is a homogeneous, folded bed, (Figure 12).

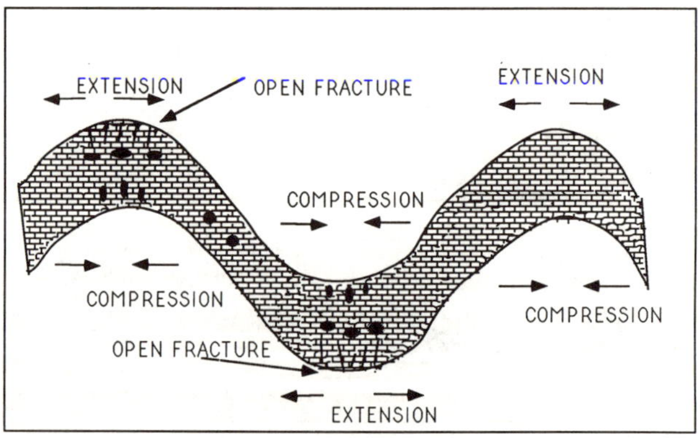

Figure 12

4.2 The implication of structural geology

All structural traps for hydrocarbons have at least one of these structural components and often a combination of them all.

4.2.1 Faults

Faults can determine whether or not a potential reservoir could hold hydrocarbons. **Normal faults** tend to be open and are caused by a system of extension in the horizontal plane. Often they are effective drains and provide links between reservoirs for fluid and gas flow, however they are not always open. A fault plane can become, due to localised pressure decrease, a site for premature or syntectonic crystallisation of minerals such as quartz, anhydrite, dolomite or calcite, none of which are very permeable. In this case, the fault acts as a barrier. **Reverse faults** often occur as barriers closing off faulted zones to general circulation. Sealing is due to the compressive nature of this fault type, although changing local conditions can alter the situation. **Tear faults** can be a barrier or a drain depending on whether crystallisation has occurred and on relative displacement. If a porous formation became (due to faulting) juxtapositioned to an impervious zone, isolation of a reservoir or pressure regime could occur.

4.2.2 Folds

Folds provide the trap into which fluids can migrate under the force of gravity.

4.2.3 Joints and fractures

Joints are important in that they can deprive an impervious rock of its ability to act as a seal. Many seals or cap rocks, however, have plastic behaviour such as clays, which means they are self-repairing. Fracture intensity depends on stress field intensity (type of tectonic activity) and the properties of the rock undergoing the deformation process.

4.3 Structures in relation to drilling practices

It is important to consider all these structures when drilling a well, as each can substantially effect the outcome of this operation.

4.3.1 Faults

Drilling personnel must take particular care when encountering faults during drilling and when tripping in and out of uncased faulted hole.

1. Faults can act as conduits for high pressure oil and gas from depth. They tend to be the cause of supercharged formations and can be extremely dangerous. If a fault of this type is encountered, a sudden influx of hydrocarbons may occur causing a kick. Often there is no warning that you are approaching a fault, so identification from seismic data is important.
2. If a fault is an impermeable, it may separate two contrasting pore pressure regimes which can cause a number of problems (see Figure 13).

(a) If the area below the fault has a considerably higher formation pressure and porosity, there is potential for a kick to take place.

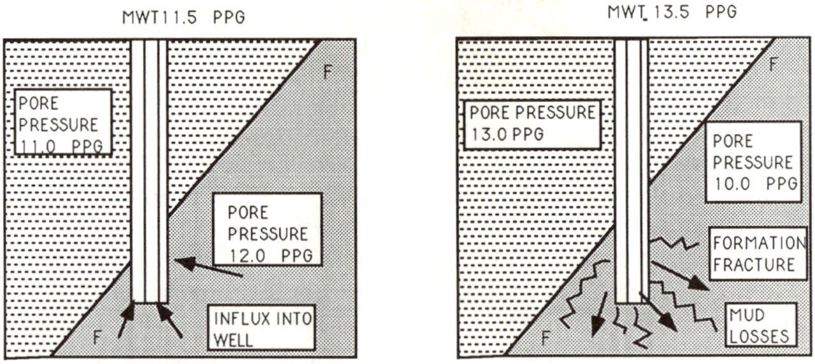

Figure 13 Influx into a well and fluid losses

(b) Crossing a fault into a lower pressured porous zone can create a number of consequences:
(i) **Fluid loss** can occur, which in turn has problems associated with loss of hydrostatic head, which cound result in loss of primary well control.
(ii) **Formation fracture** may result if the mud weight being used is greater than the formation fracture strength, leading to fluid losses and possible hole problems.
(iii) **Differential sticking** is a danger if the lower pressured zone is porous and losses occur.

(c) In relation to directional drilling, the fault plane itself, if it has a hard crystalline form, may deflect a drillstring and change the BHA's directional response. The sudden change in formation type may also affect the directional properties of a BHA. These effects are difficult to quantify as they are a result of a combination of factors which vary with each individual case.

(d) When running in and out of the hole and there is a fault in uncased open hole, care should be taken. A fault, even though it may not have affected drilling initially, is still a potential plane of weakness and decreasing and increasing relative hydrostatic head, with swabbing and surging, may open a fault resulting in losses and formation fracture.

4.3.2 Folding

A Drilling Engineer must be aware of folding structures for a number of reasons. First of all, before proceeding, the terms dip and strike off a bed should be explained, see Figure 14. Dip is the maximum inclination from the horizontal of the plane. Strike is the horizontal direction at right angles to dip.

1. **Hole stability** problems arise if the angle of dip of beds being intercepted by the well is high. Loose formations such as shales tend to cave or slide causing

holes to be unstable. The composition of the rock is therefore very important when judging hole stability (see Figure 15). Well bound homogeneous material should remain stable e.g., hard, unfractured limestone. However, loose, thinly bedded material with mineral layers such as chlorite acting as slipping planes may be highly unstable. Gravity force may overcome the internal resistance of the rock.

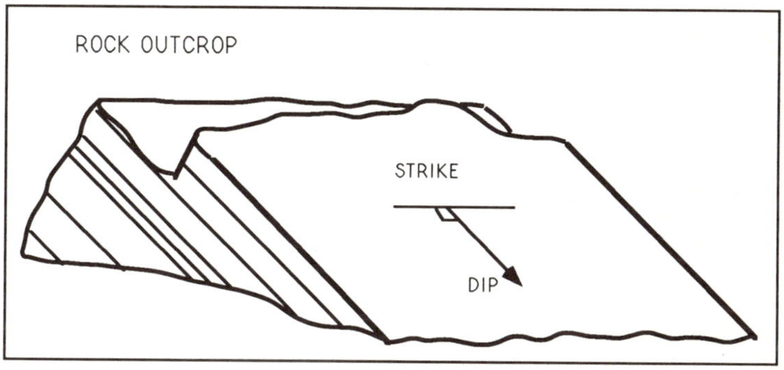

Figure 14 Dip and strike

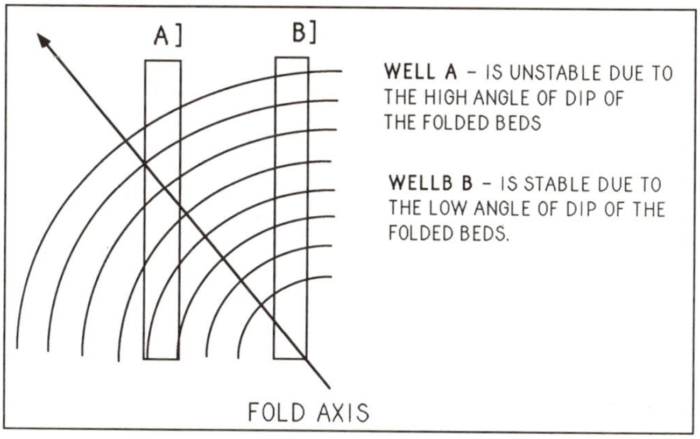

Figure 15 Well stability

2. **Directional Drilling** at right angles to the bedding plane is the ideal drilling situation, so that no deflection takes place. The dip and strike of the formation can affect the behaviour of the BHA. If a folded structure is being drilled, dip may increase or decrease with depth, so changing its effect on the BHA gradually. The relative hardness of a formation also effects directional properties as stabiliser wall contact and friction may vary dependent on formation type. Changes in formation type can give different directional responses for the same assembly.

4.3.3 Joints

Fracture joints in rocks can cause problems of losses especially in brittle rock types. Fractures however, are very important in many carbonate reservoirs, as limestone may not have any or only a little original porosity, but the volume of fractures acts as the reservoir.

5. HYDROCARBON ACCUMULATION CRITERIA

Most oil and gas in the world is found in carbonates or sandstones. However, occasionally, reservoirs which consist of shale or of fragmented basin do occur. In terms of volumes of sedimentary rocks, sandstones are more abundant than carbonates, yet more of the world's hydrocarbon reserves are in carbonates. It should be noted this figure is influenced by the volume from the Middle East, where carbonates dominate.

Most oil and gas that is produced from sandstones is derived from river borne sediments; the reservoirs often being contained within deltaic complexes. Examples of this in the Tertiary are the Mississippi in the USA, McKenzie in Canada and Alaska's Prudo Bay. Aeolian (windblown) deposits are much rarer, but can be of significant importance. A good example is the Gronigen Gas field in Holland. This is contained in the Rotliegendes formation of the Permian system which extends from NE England through the Netherlands to Germany and is a significant gas producer in the southern sector of the North Sea.

The majority of carbonate deposits are found in reefal environments. However there are a number of significant fields which are not reefal based. Chalk and dolomite reservoirs play an important part in carbonate production.

Initially there are four basic requisites for oil or gas accumulation:

1. A trap for the oil to accumulate in. These can be structured, stratigraphical or a combination of the two.
2. A reservoir rock, which has appropriate porosity and permeability to hold hydrocarbons and allow them to migrate.
3. A source of rock, a bed or beds with the right source material from which hydrocarbons can be produced.
4. An impermeable caprock, to trap the hydrocarbons and stop them migrating to surface and escaping.

Figure 16 shows some examples of types of traps.

(a) fault

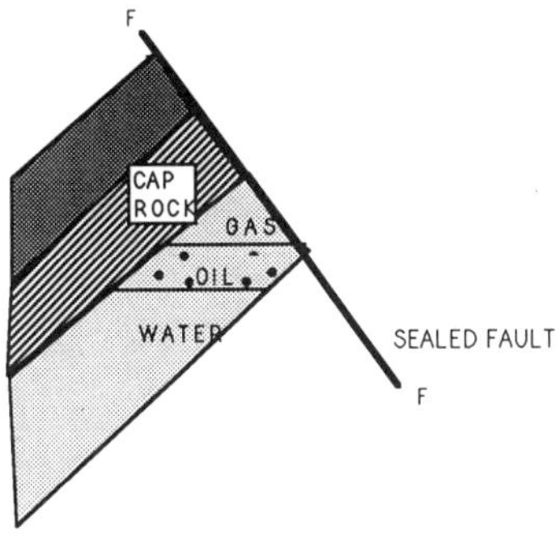

(b) unconformity

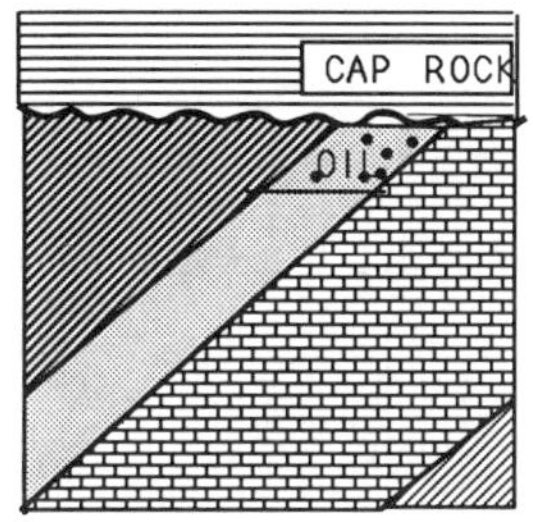

(c) salt dome

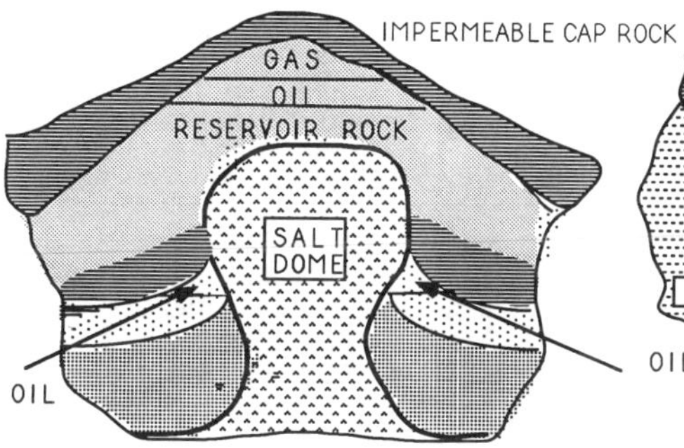

(d) stratigraphic

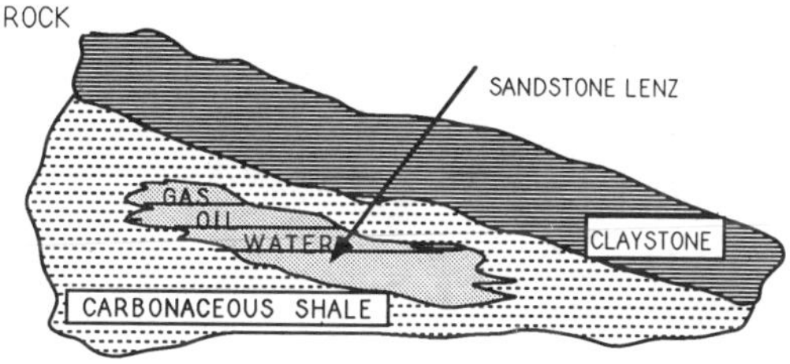

(e) reef oil

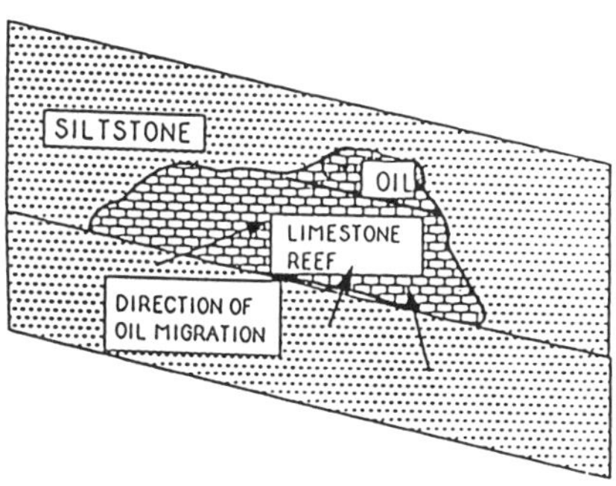

Figure 16 Well traps

All these traps have one thing in common in that they are all gravity traps with hydrocarbons migrating up into the reservoir zone were they become trapped.

5.1 Reservoir characteristics

The distribution of fluids in a reservoir rock is dependent on the densities of the fluids and the detailed capillary properties of the rock. The simplest case for fluid distribution is:

gas	top
oil	centre
water	base

Generally, there is a transition zone rather than a sharp delineation between two components within a reservoir.

5.1.1 Porosity

Porosity can be defined as the voids within a rock matrix expressed as a percentage of the total rock volume. There are two main porosity types: primary porosity and secondary porosity. Primary porosity can be defined as the porosity when the sediment was deposited. This can further be divided into intergranular and interparticle porosity. Secondary porosity develops after the deposition of sediments. The main processes of formation being solution, fracturing and dolomitisation. A generalisation that can be made about porosity in sandstones is that it tends to decreases with depth of burial.

5.1.2 Permeability

Permeability can be described as the relationship of the ease of fluid movement between interconnecting pore spaces. This is dependent on a number of factors, such as size and geometry of pores, density of fluid, viscosity, pressure and temperature conditions. The permeability of a rock can be reduced if more than one fluid is present.

It should be noted that good porosity does not automatically have good permeability associated with it. A rock may have good porosity but poor permeability, particularly in certain directions due to compaction and regrowth of minerals around grains.

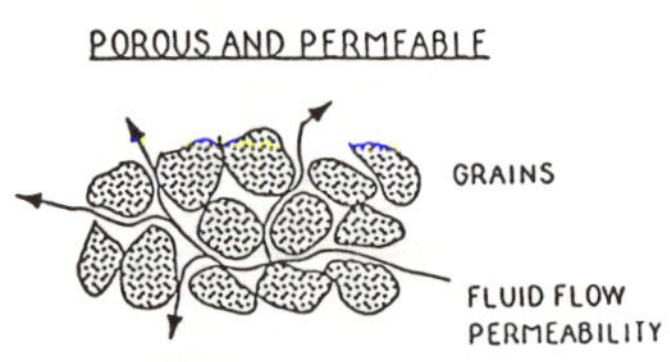

Figure 17(a) Porous and permeable

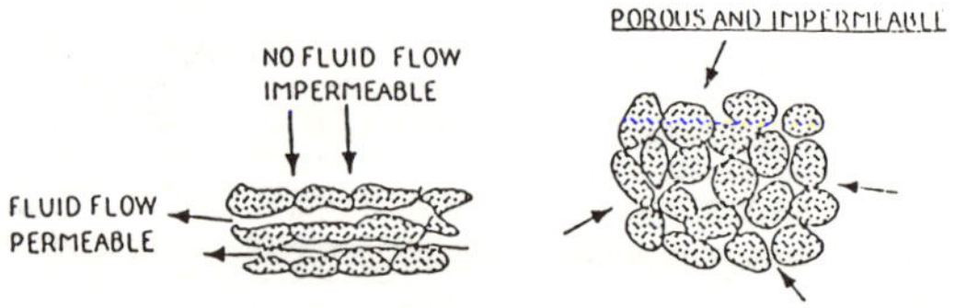

Figure 17(b) Two types of rock porous but impermeable

There are many factors which affect both permeability and porosity. These factors vary considerably between different types of rocks and even within individual reservoirs.

1. Grain size
2. Sorting of grains
3. Texture of grains
   (a) sphericity
   (b) shape
4. Amount and location of secondary minerals such as clay
5. The degree of layering of the secondary minerals in the sand
6. Cementation
   (a) type of cement e.g., calcite, silicon
   (b) extent of cementation
7. Compaction

The existence of so many factors affecting a reservoir makes the analysis a complex process and means that no two reservoirs are the same. Great variations also occur within the same reservoir making reservoir analysis complex.

6. GENERATION OF HYDROCARBONS FROM ORGANIC MATTER

After the initial burial of organic matter at shallow depth, it is broken down by the action of bacteria generating biogenic methane. With increasing depth, bacterial activity decreases gradually, giving way to chemical cracking. Cracking is the process in which heavy products (large hydrocarbon molecules) are transformed to light products (small hydrocarbon molecules). Under the influence of temperature, hydrocarbons are created from organic matter. Thermochemical generation of light hydrocarbons such as methane increases with an increase in temperature and reaches a maximum between 100 °C and 120 °C, continuing until carbonised kerogens are produced. The depth at which hydrocarbons are generated can very considerably and is related to the geothermal gradient for a region. Different areas of the world have different gradients relating to their tectonic environment.

The reason for these variations can be explained in terms of tectonic setting in relation to magmatic activity and sedimentation rates.

1. **Low geothermal gradient** In an active sedimentary basin where sedimentation rates are high and burial is fast, sediments can get buried to depth quickly and therefore are not heated to the same extent by heat conduction from depth.
2. **High geothermal gradient** In an area of magmatic activity, such as a plate margins, bodies of molten rock may be near the surface, at a depth of a few kilometers, so that the intrusion of the magma heats up the surrounding rock with heat being conducted upwards.

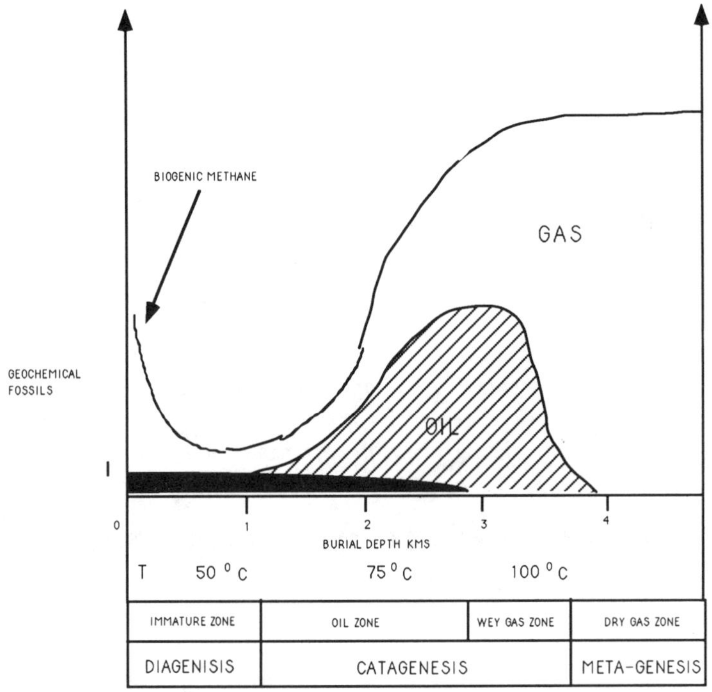

Figure 18 Formation of hydrocarbons

3. **Normal geothermal gradient** In a stable, tectonic environment the existing rock has had time to equalise temperature throughout its body and so its gradient falls somewhere in the centre of the previous two categories.

This concept of geothermal gradient is of considerable importance in the production of hydrocarbons, as the depth at which it can be produced varies with gradient. The temperature range of hydrocarbon production is referred to as a window. The composition of hydrocarbons in a reservoir affects potential productivity. The presence of other fluids, such as water (fresh or saline), gas, wet or dry, also has a bearing on well productivity.

Note: Shallow production of methane can cause considerable problems for a Drilling Engineer, as top hole drilling is often undertaken without a BOP stack. High resolution seismics (range of less than 1000 m) are used to identify shallow gas pockets of gas.

6.1 Migration of hydrocarbons

There are two basic types of migration. Primary migration is the process in which generated hydrocarbons are moved from source rocks to reservoir rocks. Secondary migration is the movement of hydrocarbons within porous and permeable reservoir beds.

The primary cause of movement of fluids from a source rock to a reservoir rock is compaction, the dewatering of sediments due to overburden pressure. Reservoirs tend to be uncompacted, whereas source rocks are compacted. This compaction corresponds to the lineation of clay minerals and/or reduction in porosity and pore fluid.

Primary migration mechanisms are a complex subject and are not within the scope of this chapter. However, secondary migration is a simpler process and is broadly argued to be due to the relative buoyancy of individual fluid components within a reservoir.

6.2 Causes of abnormal pressure

Abnormal pressure has a number of potential causes. Pressured zones have a limited lifetime dependent on the quality of the seal and the continuing existence of the reason for overpressure.

A semi-closed environment is essential for overpressure to be maintained. Rocks, however, are rarely completely impermeable and therefore pressure differentials degrade over time. Good seals for maintaining overpressure include clay and salt.

6.2.1 Overburden effect

Normally when a sediment is compacted by deep burial, fluid content and porosity is reduced. With normal sedimentation rates, expelling of fluid keeps an equilibrium with burial pressure, however, in areas of fast sedimentation, expelling of fluid may not keep up with sedimentation compression forces, causing an overpressured zone. A reduction in porosity is accompanied by an increase in bulk density. If you enter a higher pressure zone, bulk density of clays will decrease, despite consistent composition. If a clay's permeability is very low, this increases the likelihood of abnormal pressure being built up beneath, as it acts as a seal.

Pore pressure is dependent on sedimentation. Sites of rapid sedimentation such as deltas, passive continental margins etc., tend to be susceptible to high pressure. The more recent the active subsidence, the more likely abnormal pressure will be encountered. The probability of abnormal pressure occurring also increases with increased continuous thicknesses of clay. Suggestions have been made that the ratio of sand to clay in a sequence may be related to abnormal pressure magnitude. This is because sand layers may act as drains for pressure building up. The more isolated the sand bodies the less they are likely to be able to act as drains, therefore the configuration of the sediments is also a factor in abnormal pressure generation.

6.2.2 Aquathermal expansion

If a body of liquid has its temperature increased, it expands. In a sealed container, the internal pressure must increase as pressure and temperature are related. The density of the fluid will effect the pressure build-up. A sealed environment must be sealed before heating and its internal volume must be constant for pressure to build up.

6.2.3 Clay diagenesis

Unlike the release of excess water during burial and compaction, dewatering in diagnesis is the release of interlayer water from smectites (clay minerals). This dewatering is due to a combination of temperature, ionic activity and, to a lesser extent, pressure. The amount of interlayer water released is dependent on the absorption capacity of the clay minerals which in turn is dependent on their composition. This pore water can help to generate abnormal pressure.

6.2.4 Osmosis

This is defined as the spontaneous movement of water through a semi-permeable membrane, separating two solutions of different concentrations until the concentraction of each solution becomes equal, or until the development of osmotic pressure prevents further movement from the solution of a lower concentration to that of the higher concentration.

The clay layer would act as a membrane between different salinities of fluid bodies. This method, however, is thought to be restricted to a few limited number of cases and for abnormal pore pressure generation.

6.2.5 Evaporite deposits

Evaporites have two roles in pressure generation:

1. A passive role as a reservoir seal.
2. An active role in which sealed, pressured units can be transferred upward due to salt dipairism i.e., the upward movement of plastic salt under the force of gravity.

In conclusion, the identification of abnormal pressure has an important role to play in safe drilling practices. Knowing local geology, the history of deposition of an area and the criteria under which high pressure zones form, all help to identify potential drilling problems.

7. EXPLORATION TECHNIQUES

There are a number of methods of locating potential hydrocarbon reservoirs other than simply drilling random holes. These techniques can be divided into the following categories:

geophysical
correlation

7.1 Geophysical techniques

Geophysical techniques are used to establish a picture of the subsurface rocks and relate it to surface outcrop. In areas with no rock exposure, geophysical methods are often the only alternative.

7.1.1 Seismic surveys

The most common method of exploration is seismic survey. In basic terms, it employs a source that directs acoustic energy at the rock and geophones which detect the energy waves when they reach surface. There are two types, depending on the wave path taken: refraction and reflection. Reflection and refraction take place at the interface between rocks of different acoustic properties. The time taken for a seismic impulse to pass from source or shot point to the detector via the reflecting or refracting interface in both directions i.e., up and down may be used to construct a picture of geological structure at depth. Seismic reflection is the most commonly used of the two techniques. Travel time is measured in 1000ths of a second and is recorded on magnetic tape, which is subsequently data processed. The basic principle of the technique is that it shows at which depth changes in lithology occur, as seismic velocity is related to the density of the rock. This process, however, only works if two layers have different velocities. Different seismic velocities can give an idea of individual rock types.

Several types of energy sources are available for surveys. On land, a thumper which involves dropping a large weight is the most basic type. Vibroseis is used, which gives out energy as a continuous varying frequency source (usually a plate on a road surface) for 7–21 seconds. At sea, an air gun is used. A chamber charged with compressed air is then released explosively in the sea.

As sources, arrays of air guns are sufficient for petroleum exploration depths in marine operations. Marine receivers are called hydrophones. Groups of hydrophones are linked as streams 2–3 m in length and towed behind a survey vessel at a steady rate of 4–6 knots, 8–10 m below the surface. Shots are fired continuously in 10–15 second cycle intervals. Accurate vessel positioning is necessary for good data quality and this is achieved by radio and satellite navigation.

Information is presented in the form of a seismic section. Laterally equivalent events (velocity changes) show up on section. These represent reflected or refracted events and are plotted on maps. Lines joining reflectors are drawn called Isochrons, these are equal time values.

7.1.2 Gravity surveys

This technique, along with the magnetic technique, is generally used for regional rather than detailed geophysical assessment. Minute variations in the force of gravity are measured at surface by a gravimeter. These variations are caused by different densities of subsurface rock. Crystalline basement, generally, has higher densities than overlying sediments, therefore gravity surveys can be used to outline sedimentary basins.

Older, dense rocks can also be identified by this method. For example, the cores of anticlines near surface will show anomalously high readings. Salt, however, has a low density, so salt domes are easily identified with low readings.

The units of measurement are gals with the poles being 983.221 and the equator 978.047 gal. Variations of 0.0010 gal. can be important in oil exploration. Instrumentation can measure up to 0.01 milligals. An area is surveyed by intersecting traverses, generally spaced half a mile apart and readings are corrected for latitude, elevation and topography.

7.1.3 Magnetic surveys

Igneous and metamorphic rocks tend to form the basement below sedimentary basins. These rocks contain ferro-magnesium minerals and so show distictions or anomalies in the earth's magnetic field. The magnitude of the anomaly is related to the distance from the source. This can be used to deduce the thickness of sediment overlying the basement. Measurement is done from aircraft which is flown in a grid pattern similar to a gravity survey. Compensation for different types of tectonic structures are taken into consideration in calculations for different areas.

7.2 Correlation

Correlation is the use of known, existing information to predict structures in areas which have not been explored. Geological time periods can be correlated over large areas. An example of this could be the Kimmerage of the Jurassic in the North Sea. In some wells it may be at 5000 ft and in others 7500 ft deep. This implies that faulting, folding or some other geological process has occurred, either to bury, or uplift this formation between two areas.

8. THE APPLICATION OF GEOLOGICAL TECHNOLOGY FOR DRILLING ENGINEERS

Knowledge of the anticipated well geology has a major influence over the final well planning and engineering process.

8.1 Temperature gradient

Different areas of the world have different temperature gradients depending on tectonic environment. This temperature gradient combined with the prognosed depth can be used to work out approximate bottom hole and circulating mud temperatures. High pressure zones will also affect the well temperature, due to the relationship between pressure and temperature. Identification of temperature is essential with reference to selection of rig, equipment and operation, particularly on deep, high temperature wells.

All seals, elastomers etc., on surface equipment must be rated to temperature levels predicted for safe working practice. Wireline tools will be effected by temperature and may have restrictions on maximum bottom hole temperatures for operation. This should be considered, particularly if the tools are necessary for maintenance of safe practices or formations evaluation. Casing design will be affected by temperature as casing tensile and compressive strengths can vary under different temperature conditions.

In high temperature regimes the selection of mud types and chemicals must be considered carefully as mud properties may vary considerably if a large temperature range is encountered. Chemicals such as CMC are only stable below certain temperatures (250 °F). Mud salinities can change with fluid temperature variations etc. Dissolved gas is another danger in that gas may be more readily absorbed in drilling muds at high temperatures especially in oil-based mud.

8.2 Formation composition

8.2.1 Chemical composition

The chemical composition of the rock types being drilled can have implications for Drilling Engineers.

Limestones have few problems associated with them, however, calcium carbonate dissolves in water-based muds and can lead to high levels of dissolved drilling solids.
Sandstones also have few chemical problems associated with them as they are largely insoluble.
Shales the composition of shales is very important for Drilling Engineers. Different mineral compositions can have a marked affect on hole stability and types of mud systems used. Shales have a strong wetting reaction with water. When they come into contact with hydrous fluids, they absorb water and expand to many times their initial volume. Different clay minerals within shales absorb varying amounts of water, so the shale composition can be directly related to shale reactivity. Commonly occurring minerals within shales are kaolinite, illite, chlorite and montmorillinite. These are Na, K hydrous and Al silicate minerals formed from the breakdown of igneous material.
Kaolinite is a very common weathered product of feldspar in conditions where the alkalis of potassium and sodium are removed. Kaolin is common in most marine clays and becomes unstable in contact with seawater. Calcareous sediments have little or no kaolinite.
Illite is abundant in marine clays and predominates in more ancient sediments. It is stable with its non-expanding lattice.
Chlorites are decomposition products of ferro-magnesium mineral usually associated with basic igneous rocks as a sedimentary source.
Montmorillinite is the most important mineral as it can potentially multiply its volume and is very sensitive to water. The reason for this sensitivity is its large cation exchange capacity. For drilling shales with montmorillinite, inhibitors must be added to the mud to stop shale swelling.

Geologists may know the composition of shales within individual formations encountered and if there is a possibility of having reactive clays present, then preventative measures must be taken.

In anoxic reducing environments such as black carbonaceous shales, hydrogen sulphide can be formed by the action of certain bacteria. Along with hydrocarbons, it can be reservoired within porous formations. Precautions for H_2S must be taken at all times while drilling, especially on wildcat wells or in areas known to be associated with H_2S producing formations.

Chemical compositions of rocks also have implications for the erosion of drillstrings, casing and surface equipment. With corrosive drilling fluids, equipment must be closely monitored and regularly maintained and cleaned whenever possible.

8.2.2 Solid composition

Solids control planning will relate to the formation type being drilled. High sand content in drilling fluids from drilling may result in erosion of pump lines and circulating equipment. Proper selection of shaker screens, de-silters, de-sanders and centrifuges can significantly reduce use of equipment, as well as improve mud qualities. Pre-planning equipment requirements and configuration is important to cover the range of formation types expected to maximum effect.

Limestones structure can cause problems with fractured blocky limestone collapsing into the wellbore causing stuck pipe or bridging.

Sandstones Hard abrasive sandstones can wear bit gauge very quickly and as a consequence stuck pipe can result from under gauge hole. Knowledge of formations encountered can optimise bit selection in well planning.

Shales Soft shales and claystone can ball up bits and be associated with clay balls. Knowledge of this type of formation can help avoid these problems.

8.3 Seafloor stability

In many areas of the world, seafloor stability can be a problem for the positioning of rigs. Knowing the depth and type of recent sediments can help give a framework to plan and overcome problems. Geological interpretation may give an indication of the depths to which it is necessary to drive a conductor for drilling in unstable sediments. Shallow gas can also be identified by bright spots on shallow seismic survey. Initial rig selection will be influenced in some areas by seafloor conditions.

8.4 Casing and cementing

Identification of suitable rock types and depths for setting casing is necessary in the planning stage so that the appropriate amount of casing is on rig site. Good geological interpretation can reduce the stock of casing needed and provide a better seat. Fracture gradients of formation types are needed to plan the casing programme, identifying where and how many strings of casing to set. Identifying good strong casing shoe formation is a priority for well control. Knowledge of fault types and orientations also help this planning process, therefore helping to drill safer wells.

Cementing can be affected by formation chemistry (e.g., saltwater acts as an accelerator on cement), so identification of potential porous permeable sources is important. Also, different gases can affect the setting of cement. Salt formations can shear casing, so extra strength casing may be needed in areas of large-scale mobile salt accumulations.

8.5 Stuck pipe

Mechanical stuck pipe can be a problem in certain areas, particularly in thinly bedded alternative soft and hard formations. Identification of problems such as this, in certain formations, may affect the BHA choice. For directional wells, the kick-off point can be selected to avoid this problem if it is identified early enough.

Differential sticking problems tend to occur in porous permeable formations particularly sands. Selection of drilling assemblies, particularly slick ones, must therefore be carefully considered if you are going to enter a sand zone within the next bit run. Greater knowledge of the formations can help the engineer make better assembly selections.

8.6 The use of jetting techniques for direction drilling

Knowledge of formation type and the depths of changes in formation will influence the potential success of jetting in unconsolidated or soft formations. In many formations, jetting can be faster and more efficient. For small intervals of soft rock it may not be practical but for large intervals it can represent large cost savings.

8.7 Mud composition

Selection of mud type and composition must be related to predicted geology. The mud must not contaminate the formation or react with the formation, yet it must efficiently cool the bit, carry the cuttings to surface, reduce filter loss, support the weight of drill and casing string, promote maximum penetration rates, control corrosion and secure maximum hole information. A better understanding of rock type can improve decisions relating to mud composition.

8.8 During the process of drilling

The primary geological information during drilling comes from the mud logging company. Maximising the use of their geological information can enhance drilling. Description of cuttings can be very important as it often shows trends in the formation sequence. It may also give an indication of a fault being crossed or of sudden changes in formation. Bulk density can be used to predict pore pressure as density decreases with increasing pressure. The shape and size of cuttings also gives an indication of pore pressure with larger cuttings in similar formation indicating pore pressure increase. Chemical analysis shows changing clay types, giving an early indication of potential problems such as hole stability or swelling. Casing points are often picked on information given by the Loggers.

During drilling, gas analysis and trends in gas volume from the formation must be carefully observed. These observed gas levels can show changes in composition of shales, potential source rocks or reservoir rocks.

Knowledge of the formation composition and its porosity and type can be of aid to the engineer if loss circulation becomes a problem. Different porosity sizes and types demand differing responses when using loss circulation material.

Calculations such as D exponent, Sigmalog (Geoservices), Nx (Exlog), LNDR Baroid and IDEL A exponent (Anadrill) all give an indication of pore pressure increases. These methods of predicting pore pressure can all help the engineer make decisions to prevent problems.

Chapter 3

DRILLING OPERATIONS POLICIES

To carry out safe and efficient drilling operations, everyone involved must be aware of the overall game plan and rules. No programme can be effectively written or carried out until these rules and objectives have been clearly stated in a Drilling Operations Policy Document. From this Drilling Operations Policy Document, the Drilling Contract, Drilling Operations Manual and the Emergency Contingency Manual can be constructed for specific operations and from these the Drilling Programme can be written for each well.

Each operator must utilise its own experience and combine this with proven oilfield practices to consider carefully everyfully every aspect of its Drilling Operations Policy Document, as it provides the basis for all of its drilling operations. A properly written document can be of great help to small operators when trying to get drilling permission from government authorities. It provides a clear indication of the operational practices that an operator uses and shows that considerable thought has gone into these.

This chapter lays out a basic Drilling Operations Policy Document format which, from the author's experience, is workable and effective. It is by design very general and should be fine tuned for specific circumstances.

1. DRILLING OPERATIONS POLICY DOCUMENT LAYOUT

The document should be laid out in a logical manner, such as:

A Statement
B Programming policies
C Logistic policies
D Pre-spud preparations
E Rig acceptance
F Drilling operations
G Well control operations
H Testing

I Suspension and abandonment
J Safety policies
K Security policies

2. EXAMPLE OF DRILLING OPERATIONS POLICY DOCUMENT

A Statement

This Drilling Operations Policy Document represents the operating doctrine of .... Oil Company. No divergence from its policies and principles will be permitted without prior approval in writing from the Board of Directors.

Signed ....
Chief Executive

B Programming policies

B.1 The highest priority, when programming a well, is to ensure that the well can meet its objectives without risk to personnel, the drilling unit or the environment (within acceptable financial constraints).

B.2 The well must be programmed to meet with all legislative and governmental requirements in addition to internal company requirements and safety policies.

B.3 For each well, specific contingency plans will be prepared to deal with the following:

operational emergencies
emergencies caused by nature
emergencies caused by third parties

This document must meet all governmental and company standards.

B.4 Once a Drilling Programme has been approved by the Drilling Manager, it must not be changed unless approved in writing by the Drilling Manager.

B.5 All wells must be designed to satisfy all environmental requirements and to minimise the impact on the environment at all times.

B.6 **Written** permission to drill in each area should be obtained from the appropriate governmental departments and/or landowners prior to commencing operations.

B.7 Casing design must include two downhole check valves in any hydrocarbon bearing formations.

B.8 Cementation design must be such that it ensures that there is no communication between hydrocarbon bearing formations and/or between hydrocarbon bearing formations and surface.

B.9 Material specifications for tubulars etc., to be suitable for predicted well conditions (H_2S etc).

B.10 Pressure temperature ratings of equipment (BOPs etc.) will be adequate for well conditions.

B.11 The mud programme will be designed with the following priorities:

safety to personnel
minimum environmental impact
hole stability with minimum required properties
solids control requirements and rig capabilities
formation damage control
string corrosion control

B.12 The likelihood of shallow gas must be established and noted on the Drilling Programme.

B.13 Wherever possible the surface location will be moved to avoid known accumulations or indications of surface gas.

B.14 Casing tests will be restricted to the **lowest** of the following pressures:

BOP rating
wellhead rating
maximum anticipated bottomhole pressure
80 per cent of casing designed test pressure

B.15 Completion and test strings will be pressure tested to the maximum anticipated bottomhole pressure.

B.16 Open hole logging will only be carried out in wells where stable primary well control has been achieved.

B.17 Any uncertainty regarding the geological prognosis for a well must be noted on the Drilling Programme.

C Logistic policies

C.1 Only twin engined helicopters will be used over water or jungle areas.

C.2 Effective communication links between Rig Site and Base Office must be maintained. A minimum of two separate means is required for offshore or remote areas.

C.3 On offshore locations, a standby boat will be used at all times. This vessel should have direct communication links with the operations base, rig, helicopters, supply boats and emergency services (coastguard). The vessel crew must be efficient in the picking up of men in the water and the vessel must be large enough to take the entire rig complement on board.

C.4 At the start and end of a well, full inventory and stock level checks will be carried out on the rig and on any supply and stand-by vessels used during the course of the well.

C.5 All tendering for services should be carried out in a fair, unbiased manner within the guidelines of local legislation and company contract policy.

C.6 All materials purchased or rented must meet API and company design specifications.

C.7 Helicopter rescue medi & casevac services should be available on a 24-hour basis for any offshore or remote land locations.

D Pre-spud preparations

D.1 All potential offshore locations will have the following surveys carried out:

sea bed survey
shallow seismic survey
soil sample coring
tidal flow
currents
weather prediction
sea state prediction

D.2 When operating a floating rig in a new area, or under new conditions, a full riser and mooring analysis will be carried out.

D.3 On residential land sites, sleeping cabins will be placed upwind of the wellbore at a minimum distance of 200 m for normal wells and 300 m for wells where H_2S is anticipated.

D.4 The rig selected to carry out the drilling operation must be capable of doing so within its designed capabilities.

D.5 Prior to spudding, a pre-spud meeting will be held for all concerned parties and contractors to ensure that all are aware of:

well objectives
responsibilities
potential problems
contingency planning

D.6 The rig must be placed accurately by an approved surveyor at the programmed well surface co-ordinates.

D.7 On floating, anchor-moored rigs, the anchors should be pre-tensioned to 150 per cent of the expected mooring tension during acceptance.

D.8 On jack-ups, the rig should be preloaded to a weight in excess of any expected deck load during acceptance.

D.9 All potential land locations will have the following surveys carried out:

environmental survey
totco survey
access survey
noise limitation survey

D.10 In acoustically sensitive areas only, land rigs noise output should not exceed 50 dBA at 100 m from the wellhead.

E Rig acceptance

E.1 Rigs must be fully in class and able to carry out the full intended Drilling Programme without requiring withdrawal from operations for any surveys.

E.2 The mud logging unit on any well must have a direct communication link with the rig floor. All instrumentation is to be checked and calibrated on each well to the satisfaction of the Drilling Supervisor.

E.3 All drilling contractor Toolpushers and Drillers must be in possession of current valid Well Control Certificates.

E.4 A trip tank must be available, correctly calibrated for the drillpipe to be used.

E.5 A Crown-o-matic or similar brake must be fitted to the rig to prevent the travelling block impacting with the crown block.

E.6 All drillpipe must be API specification and have either:

no hard banding, or
smooth grained, fine particle, flush ground hard banding.

Drillpipe must be NDT inspected every 10 000 m of drilled depth.

E.7 All drill collars must be API specification and NDT inspected every 6000 m of drilled depth.

E.8 All elevators must be NDT inspected at rig acceptance and every six months thereafter.

E.9 All lifting equipment must be inspected according to company policy and local legislation.

E.10 Flow diverters must be of the type that have a minimum of 12 inches diameter fully opened side outlet prior to the diverter element closing.

E.11 Diverter lines should be as short and straight as possible.

E.12 Diverter lines must not pass through enclosed spaces on the rig.

E.13 Two panels for diverter operation are required, one at the Driller's console and another remote, sheltered panel.

E.14 The minimum BOP stack requirement for surface stacks is:

1 annular BOP
2 ram type BOPs

E.15 The ram type BOPs must have a pressure rating greater than the predicted bottomhole pressure.

E.16 The annular BOP will have a pressure rating which can be lower than that stated in E.15, but will be specified in each Drilling Programme.

E.17 The stack configuration will be such that there is both a pipe ram for every size of DP and tubing in the hole and a blind or shear ram.

E.18 The minimum BOP stack requirement for subsea BOPs is:

1 annular BOP (retrievable on LMRP)
4 ram type BOPs

E.19 The subsea ram type BOPs must have a pressure rating greater than the predicted bottomhole pressure.

E.20 The subsea annular BOP will have a pressure rating which can be lower than that stated in E.19, but will be specified in each Drilling Programme.

E.21 The subsea stack configuration will be such that there is both:

a pipe ram for every size of DP, and tubing to be used in the hole, and
a shear ram.

E.22 Dual fail-safe closed valves will be employed on all choke and kill line inlets and outlets at the subsea BOP stack.

E.23 All subsea ram type BOPs must be equipped with a positive locking mechanism.

E.24 The accumulator must allow two independent control systems on each BOP function.

E.25 Accumulators must be sized to allow, after a total power failure, each function to be closed, opened and then re-closed.

E.26 All elastomers must have a working temperature rating in excess of anticipated bottomhole temperatures.

E.27 The de-gasser and mud gas separator must be of a size and rating to take a pure gas kick of the influx size used in casing design calculations for the well.

F Drilling operations

F.1 The rotary table elevation, relative to ground level on land rigs and mean sea level on offshore rigs, must be established and recorded on the IADC Daily Report.

F.2 Leak-off or limit tests will be performed on each casing, other than conductors, prior to drilling each new hole section. Liner laps will be inflow tested and pressure tested.

F.3 All drilling breaks must be flow checked, fully investigated and reported.

F.4 Flow checks must be performed:

prior to POOH
at the casing shoe when POOH

F.5 If any new hole has been drilled, a full 'bottoms up' circulation must have been carried out prior to POOH.

F.6 The Drilling Supervisor will be on the rig floor during all trips out through new hole to check for hole problems and swabbing.

F.7 Any deviation from expected hole fill up volumes must be investigated.

F.8 Slow circulation rates must be taken:

once per shift
every 100 m drilled
if BHA or mud weight is changed

A minimum of two pump rates will be used for each available pump.

F.9 Choke line pressure losses must be determined and recorded on floating drilling units:

during BOP tests
if fluid properties are changed

F.10 During circulation, the shale shakers must be permanently manned and pit levels monitored constantly.

F.11 A full dimensional check of all drilling, testing or completion tools should be carried out and the results recorded prior to running tools in the hole. The record must be available at the drilling site.

F.12 Wear bushings must be installed in the wellhead during all drilling operations. The wellhead design must take this into account.

F.13 All wells will be surveyed according to specifications laid down in the Drilling Programme to meet legislative and company requirements.

F.14 All casing must be checked, drifted and measured accurately by two independent teams when it arrives at the rig site. The Drilling Supervisor is responsible for the casing tally.

F.15 Samples of bulk cement, mix water and additives should be sent to the appropriate laboratory for testing, well ahead of time, to confirm slurry properties.

F.16 A minimum stock of cement should be carried on each rig sufficient to carry out an open hole plug back job of 150 m in height.

F.17 Mud properties should be checked at least twice during each 12-hour shift.

F.18 BOP tests using water to API and company specifications should be carried out at each casing depth and more frequently if hole conditions dictate.

F.19 A ditch magnet should be utilised in the shale shaker header box at all times. It must be checked once per shift.

F.20 Casing wear must be considered in all operations and evaluated by the Drilling Supervisor on a daily basis.

F.21 On multiwell platforms, cones of uncertainty on well trajectories will be plotted on the current well's deviation plot. A Collision Monitoring Programme will be utilised.

F.22 On multiwell platforms and land cluster directional wells, kick-off points should be as shallow as possible and staggered to maximise vertical separation between adjacent wells.

F.23 Wells with the maximum horizontal displacement should be drilled from the outermost conductors on multiwell platforms or land clusters.

F. 24 On vertical wells, hole angle surveys will be taken every 100 m provided that the hole angle is below 3 °. About 3 °, magnetic single shot surveys will be utilised until the hole angle has been brought below 3 °.

F.25 Magnetic multishots should be taken at casing setting depths and at TD.

F.26 Gyro multishots should be taken in the first casing to be set over the build-up section and in the production casing.

F.27 Electric logging tools which contain a radioactive source should never be run as the first log in a logging suite.

G Well control operations

G.1 All drilling completion and workover operations will be performed with drilling fluid of such a gradient that primary control is maintained and a 'trip margin' of minimum 8 bar is incorporated.

G.2 The Drilling Supervisor must ensure that the rig crew is fully briefed on all potential well control problems.

G.3 The Drilling Supervisor is responsible for ensuring that all well control equipment on the rig is fully operational, tested to company standards and to those specified in the Drilling Programme.

G.4 Kick tolerance must be calculated on a daily basis and/or when mud weights have changed and/or weaker formations have been encountered.

G.5 A minimum stock of 100 tonnes of weighting material should be carried on the rig on any exploration well. Sufficient gel or viscofiers to weight up the mud system effectively must also be carried. On development wells, if pressures permit, the minimum stock can be reduced to 50 tonnes; below these levels, drilling must stop until stocks are replenished.

G.6 Shallow gas kicks will not be shut in. Initially, the well will be allowed to flow to deplete the shallow gas zone, whilst pumping fluid into the well at the maximum sustainable rate. In applications where a diverter system is installed, the flow will be diverted and not contained.

G.7 A near bit float valve will be run in the drillstring until the surface casing string is set.

G.8 Drills will be carried out prior to spud to ensure crews can carry out diverting procedures safely and effectively.

G.9 Wells from floaters will be drilled riserless unless:

a mud system is required to drill the surface hole for surface casing, or an evaluation of relevant factors indicates that it will be safer to drill with a riser and diverter system installed.

G.10 Wells which are drilled from floaters with a riser and diverter installed should be prepared to unlatch the riser assembly and subsequently move the rig off location as quickly as possible if the well kicks.

G.11 On moored rigs the anchor winches must be held on their brakes only to allow quick release. This will facilitate moving off location in the event of shallow gas blowouts.

G.12 On all wells where there is a possibility of surface gas, the drillstring will be pumped out of the hole until surface casing is set.

G.13 Wells will not be drilled through shallow gas seismic anomalies from jack-ups.

G.14 A diverter system will be installed during all surface hole drilling from jack-ups.

G.15 A 12 ¼ inch pilot hole will be drilled in surface sections of exploration wells.

G.16 Platforms will not be placed in areas of known shallow gas.

G.17 On platforms with injector wells, the Drilling Supervisor must monitor the activity of the injector wells in the vicinity of the drilled well on a daily basis to investigate the potential of charged up formations.

G.18 The casing string should be tested if more than 30 days have elapsed since bumping plugs, or if there is any indication of excessive casing wear.

G.19 The casing string should be tested immediately prior to drilling into known hydrocarbons if there has been any indication of casing wear.

G.20 The BOP stack and wellhead in place at any point during the course of well must be rated at maximum anticipated bottomhole pressure.

G.21 Kick-detection and shut-in drills must be held regularly.

G.22 A kick sheet must be updated daily and at every change in mud weight or BHA.

G.23 The BOP stack must be tested weekly and prior to drilling out each casing shoe.

G.24 The de-gasser will be checked daily.

G.25 The choke manifold must be fitted with a drillpipe pressure gauge at the manual choke position.

G.26 The choke pressure gauge in use should be of such a range that the maximum anticipated annular pressure in each scenario will represent around 75 per cent of instrument maximum scale reading.

G.27 Drilling spools and BOP stack side outlets must be fitted with double valves. The outer valve will be used routinely and the inner valve kept for use in case of failure of the outer valve.

G.28 Whilst running any tools on wireline inside the drillstring, a full opening inside BOP must be made up to the drillpipe sitting in the rotary table. A means of cutting the wireline must be at hand on the rig floor.

G.29 A full opening inside BOP on a counterbalanced line, with a pin connection to suit all drillpipe and drill collar sizes in use on the rig, should be installed on the rig floor.

H Testing

H.1 The Drilling Supervisor has overall responsibility for the testing operation and will check all safety and gas detection equipment prior to commencing testing operations.

H.2 A detailed programme for testing must be prepared and approved by the Drilling Manager. Any amendments or alterations to the testing programme must be approved in writing by the Drilling Manager.

H.3 A pre-test meeting must be held on site with all the relevant company and service company personnel present.

H.4 After testing, the test string must be circulated clean prior to POOH.

H.5 Open hole test packers will not be used in floating rig operations.

H.6 Gas wells should be tested with premium connection tubing.

H.7 All downhole testing and completion equipment (excepting tubulars) must be pressure tested to the maximum anticipated operating pressure prior to running into the wellbore. Tests are to be witnessed by an operator's representative.

H.8 All testing will be carried out under a double barrier principle.

H.9 Air lines to burners must be independent of rig supply and must have non-return valves fitted.

H.10 The start of a flow test should take place in daylight. Subsequent tests on the same zone may take place at any time.

I Suspension and abandonment

I.1 All wells permanently abandoned or suspended must be left such that:

legislative requirements are satisfied, and
potential leakage of formation fluid to surface and between reservoirs has been stopped.

I.2 All cement plugs sealing off hydrocarbons must be tested both by weight and pressure.

I.3 Before leaving an offshore location, a seabed survey must be carried out and the seabed left clear of debris.

I.4 Abandoned land sites must be restored to original condition with all top soil replaced unless the landowner wishes otherwise and has given instructions in writing.

I.5 Suspended land wells must be protected effectively against tampering and/or sabotage.

J Safety policies

J.1 The Chief Executive of the company is responsible to the Board of Directors for the safety of all company operations.

J.2 Every person employed on the rig should be provided with a copy of instructions to be followed in the event of an emergency. Such instructions should also be prominently displayed so as to be easily and frequently seen by all employees.

J.3 The Drilling Supervisor must keep a log of all arrivals and departures and the purpose of any visit by non-crewmembers.

J.4 **Breathing apparatus** and **medical first aid kits** must be readily available. Notices must be displayed indicating where breathing apparatus, medical first aid and fire fighting equipment are stored. The Drilling Supervisor must ensure that the appropriate crew members know the locations and use of this equipment.

J.5 Leaking hydrocarbons should be prevented. If, however, such a spill occurs, the spilled oil should be cleaned up immediately.

J.6 Paints and **flammable** materials must be isolated and stored in one place. Care must be exercised in storing mud, chemicals and cement.

J.7 The quantity of **flammable** materials, explosives and reactive, corrosive or dangerous chemicals, necessary for the operation should be kept to the minimum. These materials should be stored under suitable cover and at a point away from operations.

J.8 Chemicals, acids or other substances, which react if combined, must be kept adequately separated.

J.9 The well site should be kept neat and clean.

J.10 Floors, gangways, passages and staircases must be kept clean and free from oil, grease or other substances and clear of all obstructions.

J.11 Suitable metal bins with close, fitting lids must be provided for the disposal of cleaning materials, oily rags, cotton waste and other refuse. They must be emptied at frequent intervals and combustible materials disposed of safely.

J.12 Facilities must be provided for adequate laundering and handling of clothing.

J.13 In order to ensure safe working conditions, as much space as possible must be provided around all equipment.

J.14 **Flammable** cleaning solvents must not be used.

J.15 **Safety helmets** must be worn at all times by all persons, including visitors, on or around the rig. The wearing of suitable gloves should be encouraged and safety footwear is obligatory for all but casual visitors.

J.16 **Safety belts** must be worn by derrick men and this equipment should be inspected monthly.

J.17 The **safety belt** tail rope must be securely tied to the **derrick** or **mast** approximately three metres directly above the working platform and arranged in such a manner that there is no danger of its fouling the running gear.

J.18 For development drilling in sour crude areas and for all exploration drilling, **breathing apparatus** must be available for all members of the drilling crew.

J.19 All of those working aloft must be equipped with **safety belts** and life-lines, If constantly moving from one point to another, inertia reels should be utilised.

J.20 An escape device for the derrick man, such as an **escape line** with travelling carriage, should be provided. Any such device should be inspected and tested weekly, but should not be used except in an emergency.

J.21 Moving parts, such as chains, belts, sprockets, shafts, couplings, clutches, or any other moving part liable to cause an accident or injury, must be properly guarded.

J.22 Guards of all driving chains must be strong enough to withstand the impact of a broken chain.

J.23 The periphery of the **rotary table** must be shrouded.

J.24 The **'rathole'** and **'mousehole'** must be covered at all times when not in use.

J.25 Steel guard rails must be installed on all walkway bridges and stairways. The top rail of guards along staircases should be smooth and preferably made of tubular steel. The guard rail or fence must be at least one metre high. A toe-board must be installed.

J.26 A fixed hydrogen sulphide detector with alarm should be installed on the rig and arranged to detect H_2S concentrations in excess of 10 ppm in any enclosed space or where a leak is likely to occur.

J.27 At least six sets of compressed air breathing apparatus, of the self-contained type must be provided, together with an adequate supply of spare air bottles (at least two per set). Notices must be displayed indicating where the equipment is stored.

J.28 An automatic stopping device which enables a safe escape from **platforms**, **derricks** or **masts** and safeguards workers, who have to perform work at great heights, must be installed in any location where the conventional means of egress could be blocked in an emergency.

J.29 Approved first aid supplies and personnel transfer stretchers should be readily accessible. Notices should be displayed indicating where first aid supplies are stored.

J.30 **Fire-fighting equipment** should be available for immediate use in the event of a fire and should be maintained in first-class working order. Before spudding a new well, the Drilling Supervisor must ensure that the equipment has been recently inspected and tested by a competent person.

After using fire-fighting equipment, all equipment should be restored to first-class usable condition or replaced immediately. The full complement of equipment should be available at all times.

J.31 Personnel on site should be familiar with the location and use of fire-fighting equipment. Regular fire drills must be held.

J.32 Water and foam are both conductors of electricity and should not be used if equipment including electric cables is live. Cut off the power supply and then use an appropriate extinguishing method.

J.33 Portable fire extinguishers should be distributed throughout an installation ensuring that there is at least one unit in each accommodation unit, two units in each machinery space and one on either side of the drawworks engines. These are minimum requirements and may need to be increased when accessibility, distance and types of risks are taken into account. Extinguishers should be conspicuously mounted either immediately inside or immediately outside doors.

J.34 Safety procedures when perforating:

(a) The Drilling Supervisor is in overall charge throughout all perforating operations. Loading, arming and firing of perforating guns will be the responsibility of the Logging Engineer and he will advise when a safe approach may be made to the wellhead.

(b) As many generators as possible should remain off throughout the job.

(c) Generators should not be switched on or off during a perforating job.

(d) No use of walkie-talkie sets or intercom is allowed. Radio silence will be observed by rig, air and sea traffic and standby boat.

(e) No electric welding will be allowed at any time during a perforating job.

(f) At the judgment of the Logging Engineer, all perforating jobs should stop when an electrical storm is in the vicinity.

(g) The Logging Engineer will ensure himself, before running in, that the potential difference between cable and rotary table does not exceed 0.25 volts. The casing/rig voltage monitor should be used continuously throughout the shooting job. Should any voltage leaks occur the source must be found and shut down until either:

any loaded gun has been lowered to 100 m or deeper below the rotary table, or
any pulled gun is confirmed as shot or has been disarmed.

(h) Personnel not involved directly in the operation should keep well away from the gun, wellhead and logging cabin.
(i) Perforation will only be carried out in daylight.
(j) During perforation, no crane lifts will be permitted over the logging cable.

J.35 Safety procedures when production testing. The following are the major safety points to take into account when production testing:

(a) Sufficient space must be created on the drill floor prior to provide safe access to test equipment.
(b) Work areas around the Xmas tree and the separators should be kept clear and there must be unobstructed access to these areas at all times.
(c) When work is to be carried out on the wellhead, a suitable platform should be erected.
(d) Cranes must not operate over or in the vicinity of separators.
(e) Adequate killing fluid of the correct gradient should be available.
(f) The rig kill line should be tested and the non-return valve checked to ensure that it is not leaking.
(g) All testing and kill equipment must be tested with a pressure above the maximum pressure that can be anticipated during the operation.
(h) The flare ignition system should be checked and an emergency flare ignition system should be available.
(i) Welding must not take place during testing.
(j) The fire-fighting system should be under pressure.
(k) Gas explosion meters, a hydrogen sulphide detector and portable breathing apparatus sets must be available. All key personnel should be familiar with their operation.
Any produced gas must be checked for H_2S as soon as possible.
(l) A Production Operator will assist the Logging or Wireline Operator in opening or closing the Xmas tree valves during perforating or wireline operations.
(m) A Production Operator will be on duty at all times; from the time that the well is perforated until the test is concluded.
(n) The first perforation on any formation must be carried out in daylight.
(o) Initial opening up of the well and start up of the separator must be carried out in daylight.

K Security policies

K.1 No information gained by or for the company before, during or after drilling operations shall be divulged to any person or persons outside the company unless it is essential for these to have this information to safely and effectively provide services for the company.

K.2 No unauthorised personnel will be allowed on the rig at any time.

K.3 All holes will be treated as 'tight hole' unless specified otherwise in the Drilling Programme.

K.4 Only authorised personnel will be permitted in the mud logger's cabin.

K.5 Only company head office is permitted to make statements to the communications media.

Chapter 4

EMERGENCY CONTINGENCY PLANNING

In this chapter, we look at the function and operation of emergency contigency planning. Until recently, operators would prepare these plans solely for their own use. However, more and more governmental authories are insisting on having such plans in place prior to issuing drilling permission.

As managers of the drilling operation, we must be involved in the construction of the Emergency Contingency Plan. We know more than anyone about the technical problems and downhole pressures that the well may encounter. The exercise is useful to drilling operations staff since to prepare or approve an Emergency Contingency Plan they have to ask themselves 'what if?', for every stage of the operation to be carried out. This structured analysis can give good technical results as a spin-off from the Contingency Planning Assessment.

In certain areas there are drilling windows which should be observed in deference to local weather and climatic conditions. It makes sense to avoid icebergs by drilling when none exist and similarly, to drill in typhoon areas during the time of the year when, statistically, typhoons are unlikely. By quantifying risks such as these at the initial drilling stage, we can provide useful information for the Production Engineers to consider when planning the production systems to be used if and when the well is commercial.

In this chapter, we break down contingency planning into its component parts, namely:

Contingency plan objectives
Classification and origin of emergencies
Protection, prevention and preparation
Manual preparation
Organisational requirements
Contingency actions
Auditing effectiveness

1. CONTINGENCY PLAN OBJECTIVES

The primary objective of an Emergency Contingency Plan is to protect life and prevent injuries.

Initial considerations are:

To establish the best timing for a specific drilling project.
To specify the potential emergencies which can be anticipated during a specific drilling project or operation.
To provide a clearly understandable list of all actions which must be taken in an emergency.
To create a working organisational framework that will guarantee a rapid and effective response to an emergency.
To designate the persons responsible for taking action in emergencies and to provide backup alternatives.
To establish and provide the communication network required to deal with emergencies.
To specify any special equipment, materials and personnel required to handle an emergency and to list their sources and availability.

Secondary considerations are:

Minimising damage to equipment and the environment.
Satisfying all company, industry and legislative requirements.

2. CLASSIFICATION AND ORIGIN OF EMERGENCIES

Emergencies can be subdivided into three main groups:

1. **Major** – the operator is unable to restore the situation to normality without external assistance.
2. **Serious** – there is significant risk to life, equipment or the environment.
3. **Minor** – there is no danger to life and only minimal risk to equipment or the environment.

The causes for the occurrence of emergencies can be divided into three main groups:

1. Emergencies caused by operational failures resulting in:
   fire and explosion
   well control problems and blowouts
   spills
   release of dangerous liquids, gases or radio-active substances
   ballasting problems
   rig evacuation
   accidents involving death, serious injury or major damage to equipment or the environment

2. Emergencies caused by nature:
high winds and waves
storms, typhoons, cyclones
forest fires
lightning
earthquakes
land slides
excessive rain and flooding
mist and fog
icing, snow and icebergs
tsunami

3. Emergencies caused by third parties acts:
bomb threats
terrorism
war zones
strikes
malicious damage
collision, impact by boats

Most drilling projects will have to consider all of the above listed operational emergencies, so each listed emergency source should be built into the Emergency Contingency Plan.

The likelihood of each natural emergency type will depend on the drilling location and the time of year. In the tropics, for example it is possible to avoid typhoons by timing the drilling operation outside of the typhoon season. Simple examples such as this illustrate the need for us, as planners, to consider the Emergency Contingency Plan prior to selecting final drilling location and timing.

By consulting meteorological records and talking to operators who have worked in the area in the past, a picture can be built up of the possible and probable natural emergencies which might have to be dealt with during the drilling project. Each of the above-mentioned natural emergency sources should be studied and either discounted as unlikely, or built into the Emergency Contingency Plan.

Similarly, third party acts can be predicted to a great extent by virtue of the chosen drilling location. These acts should be considered individually and either discarded or built into the plan as above.

3. PROTECTION, PREVENTION AND PREPARATION

It is essential that prior to embarking on a drilling project the following review of operations is made:

1. Potential emergencies that could be encountered during the drilling project.
2. Protection requirements to limit the consequences of emergencies.

3. Preparedness for dealing effectively with emergencies.
4. Whether well timing could be altered to reduce exposure to emergencies.

In many cases, natural and man-made barriers can give passive protection against the effects of emergencies, for example, restricting entry and compartmentalising can reduce the domino effect of many emergencies. Additional protection can be added in the form of deluge systems and fire mains.

Operational procedures should be set up so that emergencies can be anticipated and detected before they become major incidents. The tools at our disposal to this end include:

1. HAZOP studies and HAZARD Analysis.
2. Equipment inspections to ensure that it meets legislative and industry standards.
3. Effective detection and control systems to activate alarms, shut-downs, deluge systems etc.

Preparation is done by running through potential emergencies on paper and trouble-shooting the procedures which would be applied in a real situation. By carrying out this exercise the emergency situation needs will be confirmed and systems put in place to meet them. Drills should be carried out to test the effectiveness of the system and based on these, the systems fine timed for optimum effectiveness.

Well timing cannot be overstressed. For a single well Drilling Programme the operator has considerable flexibility in timing his well. The likelihood of some emergencies caused by nature can be substantially reduced just by delaying or bringing forward spud date by a few months.

4. MANUAL PREPARATION

An Emergency Contingency Manual should be prepared well in advance of operations, containing all the information required for those involved to understand their roles and functions in any anticipated emergency. In most drilling operations, this can all be contained within one manual, however, where drilling is being carried on in areas of knock-on effect, such as platforms, several manuals may be required to cover all eventualities.

To be effective, the manuals must contain the following information:

1. Emergency strategy
2. Personnel role designation with contact numbers and backup staff
3. Organisational charts showing communication lines and means
4. Written procedures and instructions for each anticipated emergency
5. Maps and charts with zoning and access routes
6. Lists of company and available third party stocks
7. Lists of authorities and public services contacts
8. Lists of third party contact personnel

9. Lists of medical services and backup
10. Lists of on-site safety facilities

5. ORGANISATIONAL REQUIREMENTS

The initial countermeasures in the event of an emergency must in the first instance, be taken care of by personnel who are normally in charge of the operations at the site where the emergency has developed. Therefore, these personnel should be qualified and trained to carry out the necessary first actions as outlined in the plan. When the emergency situation cannot be controlled by the initial countermeasures, a team of the Emergency Organisation, specially created for the purpose, should be activated for immediate assistance and/or take over.
The operator's Emergency Organisation must be designed to meet the requirements of the various possible emergencies.

5.1 Initial countermeasures at place of emergency

Normally, in any emergency, the first actions on site will be carried out by the person discovering the emergency followed by further decisions of the Site Supervisor once he has been contacted. The Site Supervisor will:

1. Raise the alarm
2. Ensure that the well is made safe
3. Move in fire-fighting teams or rescue teams as required
4. Inform Base about the situation as understood, with a description of all action taken to date and the likely source and effect of the emergency
5. Request assistance if required

5.2 Base emergency services

On request for assistance in dealing with an emergency, the rapid mobilisation of the base emergency services must be implemented without delay and a pre-designated Emergency Action Team mobilised under the direction of the Emergency Contingency Manager. The Emergency Action Team will organise the Base activities in the following:

1. Support services
2. Public relations
3. Press statements
4. Personnel enquiries

5.3 Emergency control centre (ECC)

An area known as the Emergency Control Centre should be designated at the base. Within this ECC, the Emergency Contingency Manager and the Emergency Action Team have a central point to organise the emergency support services. It should therefore have effective communication lines in and out and have space for secretarial and refreshment facilities.

5.4 Emergency operation centre (EOC)

The installation of an Emergency Operation Centre near or at the scene of the emergency must be given consideration for each emergency scenario. Suitable EOCs on land would be, typically, trailer-mounted units and offshore would be control centres on nearby rigs, platforms or boats.

5.5 Communications

There must be a totally reliable means of communication between all components in the emergency contingency handling operation. In practice, this means having several phone, fax, radio and telex lines with considerable redundancy in the system.

5.6 External support services

Whenever possible, emergencies should be dealt with internally by the operating company. However, government agencies and other authorities should be notified of abnormal situations which could develop into major or serious emergencies. Details of the responsibilities and capabilities of the support services which may be called upon to assist should be included in an emergency plan along with 24-hour contact numbers. A list of external authorities and support services should be built into an emergency plan, including, as applicable:

1. Police
2. Government agencies
3. Service companies
4. Medical services
5. Fire services
6. Air transport
7. Marine transport
8. Diving services
9. Oil spill specialists
10. Coast guard services
11. Search and rescue
12. Other oil companies (mutual assistance)

6. CONTINGENCY ACTIONS

As we discussed in section 2, all emergencies will result from one or more of the following:

operational problems
natural causes
third party action

In this section, we discuss the action that must be taken to deal with these problems. Clearly each well or drilling programme must be dealt with individually, however, the follow'ng provides a guide for input into Emergency Contingency Manuals.

6.1 Operational problems

These result from human, equipment or system failure and include such emergencies as:

1. Well control problems
2. Oil spills
3. Release of dangerous substances
4. Fire and explosions
5. Serious or fatal accidents

6.1.1 Well control problems

Most well control problems will be solved by the rig supervisory staff. However, sometimes matters do get out of control and a blowout follows. The effects of a blowout vary greatly and the Contingency Plan must consider each well's situation with respect to its potential flow and its environment. The factors to be considered in each case are:

1. Potential well flow
2. Potential duration of well flow prior to bridging
3. Onshore: site locations, remoteness or proximity to industry or population centres
4. Offshore: knock-on effect on multiwell platform
5. Radiation effects from blowout fire
6. Potential of toxic gas production
7. Potential of oil spill
8. Effects of smoke

Blowout control techniques Once a blow-out has occurred there is a choice of four approaches to handle the problem:

1. Do nothing and hope that the excessive production rate will allow the formation to bridge with sand production or collapse.
2. Well capping which is only possible where the wellhead area is relatively undamaged so that auxiliary spools can be placed on the wellhead or suitable casings.
3. Relief well dynamic kills – relief wells are used to enter the producing wellbore and a high rate of circulation down the relief well and up the blowout well creates sufficient back pressure on the formation to initially reduce flow and ultimately kill the well.
4. Well saturation methods – relief wells are used to saturate the producing reservoirs around the production point in the wellbore to choke off production.

Of these 1. and 2. are the cheapest options. However, since their chances of success can vary, it is expedient to start a relief well immediately after a blowout to ensure that if options 3. and 4. are required there is minimal delay in achieving these.

Relief well planning Prior to programming the relief well, the following information should be collated:

1. Positional data of nearby wells and wells being drilled
2. Predicted new formation pressure regimes
3. Casing design criteria
4. Flow area requirements in wellhead and tubing for high volume injection and killing operations

Relief well siting The following criteria should be considered when selecting a suitable relief well site:

1. Subsurface position of well(s)
2. Distance from blowout at surface
3. Access roads suitability
4. Prevailing wind direction
5. Water supply possibilities
6. Seabed conditions and obstructions
7. Water currents
8. Nearby shipping lanes

On land relief wells, the killing pump station will be located at the relief well site, consequently any site should be large enough to contain this and a vast amount of drilling fluid storage capacity.

Regaining control of well All initial procedures taking place on site to regain control must be automatic and be laid down clearly in the Emergency Contingency Manual. They should be implemented whilst simultaneous support is being arranged. These actions will typically include:

1. At site
   (a) Raise alarm
   (b) Appraise situation and inform Base together with request for assistance
   (c) Attempt to contain hazard as much as possible
   (d) Secure other wells by closing subsurface safety valves
   (e) If possible, secure and protect the installation and complement by some or all of the following:
   Shut down ancillary machinery
   Shut down process equipment
   Isolate and depressurise process system
   Reduce/remove flammable materials

Isolate and protect non-affected areas and equipment
Control access to site
Muster personnel
Deploy fire control teams
Deploy rescue teams
Prepare to abandon rig/area
Prepare to move a floating drilling unit off location
Abandon installation

2. At base
(a) Mobilise Emergency Action Team
(b) Arranging immediate assistance for fire-fighting and rescue
(c) Mobilise relief action
(d) Notify appropriate authorities
(e) Investigate methods for controlling well and containing situation
(f) Prepare relief well plan
(g) If required, organise relief well rig/site prep/logistics

6.1.2 Oil spills

In establishing their anti-pollution plans, operators should study their operating methods and conduct surveys of the regions to identify vulnerable areas and areas of ecological importance. Any risk areas highlighted by the surveys form the basis for contingency planning. Emphasis should be placed upon the requirement to provide protection and containment. Subsequent collection and removal measures are largely remedial, totally dependent on the particular situation and may involve third parties.

In the event of an incident, steps must be taken immediately to assess the situation, eliminate the sources and contain the pollutant. The appropriate local authorities should be consulted when deciding upon removal and clean-up methods, as their approval may be needed.

Prevention Working practices should be such as to minimise the likelihood of accidental release. Drip trays and closed drain systems should be designed to contain likely spillages prior to detection and treatment. In areas of ecological sensitivity, protective booms and skimmers must be available.

Collection and containment The plans used for this must be realistic for the prevailing wind and wave conditions that could be expected in the areas. The effective methods to be used are very dependent on the wind and waves and include barriers, booms or chemicals offshore and physical damming onshore.

Removal The removal method selected is also dependent on the local conditions and the spill to be removed. The use of absorbent materials, skimming and pumping methods, chemical removal or even burning (if permitted) should be considered.

Dispersal Dispersants are now widely used in the removal of oil spills offshore and can be effectively applied in rough seas. None of the above-mentioned removal methods, with the exception of burning (which has to be done very quickly after the spill has occurred) is effective in rough seas, so in most heavy weather offshore locations, dispersal will probably be the best method to apply to the spill.

6.1.3 Release of dangerous substances

In developing contingency plans for accidental releases of hazardous materials, operators must evaluate the potential hazards associated with such products. They should identify vulnerable areas and determine the measures necessary to effect rapid containment and reduce the exposure risk to life and property.

Should an incident occur, prompt action must be taken to stem the flow at source, thereby preventing further spread of the substance. Particular attention must be paid to the prevailing wind strength and direction and the effect this may have on the spread of dangerous liquids or gases.

As an aid to determining different response levels to a particular situation, various operating conditions can be considered. Certain conditions represent situations of minor or moderate concern, whilst others may involve considerable loss of control, possibly with the release of lethal concentrations of gas. This condition, mainly applicable to blowout situations, might require ignition. This will, undoubtedly complicate subsequent control operations, but if lives are at risk, the decision should not be delayed.

For wells containing H_2S, drills must be carried out on a regular basis on the installation to ensure that crews are familiar with H_2S procedures. If a blowout occurs, consideration must be given to igniting all produced gas as quickly as possible. This matter must be covered in the Emergency Contingency Manual and not just left until the situation arises.

When H_2S or SO_2 (which is created after igniting H_2S) is released in an uncontrolled manner, it can affect areas much larger than the immediate installation. Prevailing winds can carry the gas to population centres and the possibility of this must be analysed and planned for in the Emergency Contingency Manual.

6.1.4 Fire and explosions

In the event of a major fire or explosion, the Emergency Contingency Manager will be confronted with a very confused situation. Reports will be coming in from several sources and in practice they will often contradict each other. Consequently, the Emergency Contingency Manual must provide guidelines for the Site Supervisor to send a clear and concise appraisal of the situation. The initial situation appraisal must determine the following:

1. Likely source of fire and containment potential
2. Type of combustibles present
3. Any casualties
4. Any missing crew
5. Potential knock-on effect
6. Emergency system status

Immediate priorities for the Site Supervisor are:

1. Give medical treatment to any casualties
2. Rescue any missing personnel
3. Contain or restore control to situation

To contain and control the situation the following methods can be used:

1. Stop the flow of combustibles and blow down any pressurised process facilities
2. Isolate any redundant electric circuits
3. Shut down ventilation in affected areas
4. Apply maximum water cooling to affected areas

All personnel on the installation must be conversant with all emergency procedures, which should be along the lines of:

1. Sound the alarm
2. Ensure their escape routes are clear
3. Report the situation to the Site Supervisor
4. Help any casualties to clear the affected areas

The Site Supervisor upon being notified of the fire should:

1. Direct actions, including fire and rescue teams and firefighting systems, to control and contain the emergency
2. On offshore rigs, use the radio operator to inform the standby boat, any shipping or helicopters in the area and the coastguard
3. Direct medic, HLO, waking team and life boat crews
4. Contact the Emergency Contingency Manager giving full details of the emergency and initial back-up requirements
5. Consider evacuation of the installation
6. Make a head count

6.1.5 Serious or fatal accidents

In the event of a serious or fatal accident, the Site Supervisor has three functions to fulfil:

1. To ensure that the casualty is given immediate treatment
2. To ensure that the workplace is safe and that the accident count cannot recur
3. To comply with all company and national authorities requirements with regard to incident reporting and investigation

In most cases, work at the scene of a fatality may not be resumed except with permission of the appropriate national authorities; unless, in the opinion of the Site Supervisor, to fail to do so might endanger the lives of others.

The Emergency Contingency Manual should lay down the correct formula to be employed when informing the relatives of a casualty.

6.2 Natural emergencies

As with all emergencies our priorities for action are:

1. The safety of personnel in the affected area or onboard the installation
2. Minimising damage to equipment
3. Minimising loss in production

For all natural emergencies, the Emergency Contingency Manual should detail the following:

1. The influence each emergency can have on the operations
2. How weather movements and changes will be effectively monitored
3. An effective means of evacuating installation personnel to safety

In all cases, the decision to evacuate the installation should be made by the Installation Supervisor.

6.3 Emergencies caused by third party acts

Deliberate acts of sabotage by determined individuals are difficult or impossible to prevent in most cases. The role of the Emergency Contingency Plan is to minimise damage and confusion should such an event take place. To create a meaningful reaction the Emergency Contingency Plan must consider and address the following points:

1. Potential threats
2. Vulnerable points
3. Potential support from Police and Army
4. Site security and access arrangements
5. Site lighting arrangements

Tc be prepared for bomb threats, the following organisation should be in place:

1. Effective search procedures
2. Effective bomb disposal procedures in liaison with local Police or Army
3. Established procedures for staff receiving bomb threat calls to record the maximum amount of information

7. AUDITING EFFECTIVENESS

Having established an Emergency Contingency Plan, the next step is to check its effectiveness. In practice, training has to be carried out for all personnel involved until they are fully conversant with the workings of the plan and efficient in carrying out its procedures. Training should be carried out in three formats:

1. Training sessions
2. Drills
3. Simulations

Training sessions should be used to familiarise staff with emergency equipment and procedures and their individual role in each type of emergency.
Drills should be used to check the response time of involved personnel to muster, form search teams or evacuate the installation. These should be carried out until response is satisfactory, after which they should be applied randomly.
Simulations will involve creating a simulated emergency and carrying out a dry run involving some or all of the staff to check that the total system is functioning.

Chapter 5

DRILLING ECONOMICS

To optimise drilling operations, we have to specify the yardstick by which performance is measured. The most relevant yardstick is **cost per metre or foot drilled**. Overall cost must be looked at since individual costs can be misleading. The rig operating rate represents only a fraction of the overall cost, therefore a cheap rig day rate does not always coincide with a cheap well.

Holes are drilled in the ground to provide information (in the form of cores or logs or test information), to provide production of oil and gas or to provide an injection point into a reservoir. These objectives of the well should never be forgotten during the drilling operation. To optimise drilling economics, we must achieve the objectives of the well as economically as possible. To do this, we must understand the cost allocations and proportions in drilling operations and use our technology to fine-tune these to reduce expenditure without affecting safety or efficiency.

This chapter looks at these cost allocations, discusses exactly who is controlling which costs and how they can be minimised.

1. COST SPECIFICATIONS

Drilling costs can be broken down into three groups:

fixed
daily
unit

1.1 Fixed costs

Fixed costs are those which are determined mainly by the nature of the well and include the following:

wellheads
site preparation
casing, cement, tubing and packers

Effecting economies in fixed costs, therefore, is the direct responsibility of the Drilling Manager and the Drilling Engineers, who planned the well. The Drilling Supervisor has little impact on these.

1.2 Daily costs

Daily costs are related to the time spent on the operation. On offshore rigs, they are usually the largest items of expenses and are listed below:

payments to drilling contractors (rig time)
tool rental
payment to specialist services
salaries, wages etc.
fuel
lubricating oil, grease
drilling consumables (rope, soap and dope)
transport of materials

The Drilling Supervisor on site, the Drilling Manager and Drilling Engineers can all have an effect on daily costs.

1.3 Unit costs

This is the price of a unit or a commodity such as the price per tonne of baryte or bentonite. This can usually be optimised in the tendering process, which is usually the responsibility of the Drilling Manager. Furthermore, good site supervision can ensure that consumption is not excessive.

Caution: The total daily cost rate must not be considered in isolation when comparing operations. The speed, efficiency and end product of the well must also be considered.

2. COST BREAKDOWN OF DRILLING OPERATIONS

At present, as the oil industry is coming out of recession, the costs for individual types of rig is varying considerably. The table below illustrates the average cost comparisons between rig types at time of writing:

Rig type	Total daily drilling costs ($/day)
Land rig (shallow)	15 000
Land rig (deep)	25 000
Platform rig	50 000
Jack-up rig	95 000
Semisubmersible	75 000

These figures are general and should only be used as a guide. Even though jack-up costs have been traditionally cheaper than those of semis, demand for deep water jack-ups has pushed their day rates above most semis.

2.1 Specific cost breakdown – offshore exploration wells

To quantify the costs, we must look at real well expenditure. Below is quoted a typical cost breakdown for a 1990 UK North Sea exploration well. This is based on a TDAH of 3500 m with 7 inch casing to TD and includes four days of coring and four days of testing. The total time spent on the well is 60 days.

Cost group	Cost in US $ (thousands)	% of well cost
Location survey	160	3.0
Rig mob/demob	270	5.1
Rig positioning	8	0.2
Casing	570	10.8
Wellheads	180	3.4
Rig costs	1400	26.6
Drilling equipment rental	50	0.9
Fishing tools	9	0.2
Drill bits	140	2.7
Mud	220	4.2
Cementing	170	3.2
Electric logging	320	6.1
MWD	14	0.3
Mud logging	160	3.0
Coring	60	1.1
Directional control	240	4.5
Supply boats	370	7.0
Standby boats	160	3.0
Helicopters	212	4.0
Diving/ROV	130	2.5
Weather forecasting	4	0.1
Medical services	3	0.1
Testing equipment	100	1.9
Storage/onshore transport	26	0.5
Contract staff	250	4.8
Base office	41	0.8
	5267	100

2.1.1 Discussion

In this cost breakdown, there are 26 groups. Four of these, Rig Positioning, Fishing Tools, Weather Forecasting and Medical Services, all cost less than $10 000 for the well, so fine-tuning these services will provide us with minimal savings. Our attention must turn to the remaining 22 large items where a 10 per cent saving on individual costs can substantially reduce overall well costs.

Listing the remaining groups in order of either fixed, daily or unit, we achieve a spread as follows:

Fixed item	**Cost in US $ (thousands)**	**% of well cost**
Location survey	160	4.0
Rig mob/demob	270	5.1
Casing	570	10.9
Wellheads	180	3.4
Drill bits	140	2.7
Cementing	170	3.2
Electric logging	320	6.1
Coring	60	1.1
Testing equipment	100	1.9
Fixed item total	1970	38

Daily item	**Cost in US $ (thousands)**	**% of well cost**
Rig (56 days @ 25 000)	1400	26.6
Drilling equipment rental	50	0.9
Mud logging	160	3.0
Directional control	240	4.5
Supply boats	370	7.0
Standby boats	160	3.0
Helicopters	212	4.0
Diving/ROV	130	2.5
Storage onshore transport	26	0.5
Contract staff	250	4.8
Base office O/H	41	0.8
MWD	14	0.3
Daily item total	3053	58

From this figure and knowing that the well lasted 60 days, we can calculate the average daily cost.

$$\text{Average daily cost} = \frac{\$3\ 053\ 000}{60} = 50\ 000/\text{day}$$

Unit Item	Cost in US $ (thousands)	$ of well cost
Mud	220	4.2
Unit item total	220	4

To summarise these groups, we find the following:

fixed cost items	$1 970 000
daily cost items	$3 053 000
unit cost items	$ 220 000

What this means in real terms is that saving a day on the well will save 1/60 of $3 053 000 and not 1/60 of the overall well cost of $5 267 000. It also means that an extra day spent on tripping, directional correction, treating the mud, waiting on weather and making spurious trips will cost the operator a minimum of $50 000.

In chapter 11, drilling optimisation is discussed; this, essentially involves untilising experience and data to employ the best tools most efficiently to reduce drilling costs and time spent on a well. To keep drilling costs down, a team effort is required with innovation and experience used at every decision point.

2.2 Specific cost breakdown – onshore exploration wells

For comparison with the offshore well discussed in section 2.1, the following represents a shallow (1000 m) European land well to be drilled in 1990.

The cost breakdown is for a 1000 m well with 7 inch casing set at 500 m and 6 inch hole drilled to TD. It is assumed that there are two days spent coring, the well is then logged and abandoned. Estimated days on well is a total of 20.

Cost group	Cost in US $ (thousands)	% of total cost
Site survey, rental, preparation and restoration	72	11.7
Rig mob/demob	46	7.5
Rig date rate (20 days)	97	15.8
Wellhead	26	4.2
Conductor and casing	40	6.5
Rock bits	17	2.7
Core heads service	13	2.1
Mud chemicals	14	2.3
Mud engineering	10	1.6
Mud logging	30	4.9
Cementation	35	5.7
Fuel	10	1.6
Tool rentals	5	0.8
Transport/cranes	10	1.6
Electric logging	30	4.9
Disposal	8	1.3
Accommodation	5	0.8
Water supply	5	0.8
Well engineering	50	8.1
Site supervision	15	2.4
Site geologist	8	1.3
Overheads	70	11.4
Total	616	100%

Breaking down these costs into fixed, daily and unit costs, we find the following:

Fixed item	Cost in US $ (thousands)	% of total cost
Site survey, rental and restoration	72	11.7
Rig mob/demob	46	7.5
Wellhead	26	4.2
Conductor and casing	40	6.5
Rock bits	17	2.7
Core heads and service	13	2.1
Cementation	35	5.7
Transport/cranes	10	1.6
Electric logging	30	4.9
Disposal	8	1.3
Accommodation	5	0.8
Well engineering	50	8.1
Total	352	57.1

Daily item	Cost in US $ (thousands)	% of total cost
Rig day rate (20 days)	97	15.8
Mud engineering	10	1.6
Mud logging	30	4.9
Fuel	10	1.6
Tool rentals	5	0.8
Water supply	5	0.8
Site supervision	15	2.4
Site Geologist	8	1.3
Overheads	70	11.4
Totals	250	40.6

$$\text{Total daily cost} = \frac{250\,000}{20} = 12\,500/\text{day}$$

Unit item	Cost in US $ (thousands)	% of total cost
Mud chemicals	14	2.3
Unit item total	14	2

2.2.1 Discussion

The main cost groups (over 4 per cent of well cost) for this well are as follows:

	Cost in US $ (thousands)	% of total costs
Site survey, rental, preparation and restoration	72	11.7
Rig mob/demob	46	7.5
Rig day rate	97	15.8
Wellhead	26	4.2
Conductor and casing	40	6.5
Mud logging	30	4.9
Cementation	35	5.7
Electric logging	30	4.9
Well engineering	50	8.1
Overheads	70	11.4
	496	80.7

As can be seen above, ten cost centres account for over 80 per cent of the total well cost. Looking at each individually:

Site survey, rental, preparation and restoration $72 000 11.7 per cent
This expenditure is usually under the control of the Drilling Manager and/or the Engineers who select the final site location and negotiate the deal with the land owner and site building company.

In areas of extreme environmental sensitivity, this cost centre could represent up to 25 per cent of the well cost if special restrictions were placed on site screening and proximity to existing buildings. Site access roads can cost as much as the site itself and should, in principle, be kept as short as possible or avoided altogether. Investigative engineering techniques should be employed when using any existing hard standing (e.g. old factory sites) and looking for natural screening and level areas.

Sites can be rented or purchased and in some cases, it is cheaper to purchase the site. The added advantage of doing so is that it ensures that there will be no problem with the land owner during construction and drilling. There are no hard and fast rules on site selection and it is up to the skill of the Drilling Manager to minimise these costs.

Rig mob/demob $46 000 7.5 per cent
This charge is made by the Drilling Contractor to cover the costs of getting his equipment to the site, rigging it up, and once the well is finished, returning it to its base. This is also a cost centre which is under the direction of the Drilling Manager. If suitable rigs are in the area of interest at a certain time of the year, the mob/demob costs can be reduced substantially during these periods. In order to establish rig movements it is necessary to maintain a dialogue with other operators and with drilling contractors.

Selection of smaller rigs which involves less truck movements between sites can reduce mob/demob costs. There is, of course, a limit to the smallness of the rig which can be used effectively on a given well so a compromise must be arrived at between cost and rig size.

Rig day rate $97 000 15.8 per cent
The above remarks for rig mob/demob also apply to day rates. The time spent on a well depends on the skill of the Drilling Manager, the Drilling Engineer and the Drilling Supervisor.

Wellhead $40 000 6.5 per cent
Wellhead selection is usually done by the Drilling Engineer or Drilling Manager. Company policy may dictate a certain brand or type. Otherwise tendering the various wellhead companies can result in a bargain. Furthermore, wellheads can be reused and refurbished. Wellheads of known history provide a low-cost alternative to a new one for every well.

Conductor and casing $40 000 6.5 per cent
These costs are dictated by the casing design for the well as well as local supply and demand conditions. The Drilling Engineer who programmes the casing design can

provide several alternatives which meet the well needs. The alternatives can then be costed out and the cheapest one selected.

Mud logging $30 000 4.9 per cent

Selection of the most suitable mud logging company is by a tendering exercise controlled by the Drilling Manager and Drilling Engineer. The Drilling Supervisor can influence overall mud logging costs by keeping the days spent on the well to a minimum.

Cementation $35 000 5.7 per cent

When the well is programmed by the Drilling Engineer, certain basic cementation requirements will be established. The traditional cementation service companies all operate at similar costs, so the tendering exercise for services will produce the cheapest bid. If the Drilling Manager is prepared to use premixed slurry for cementation in areas where this is readily available, and use a simple displacement pump to put it in place, then considerable cost savings are possible. The Drilling Supervisor can assist cost control on cementation by accurately predicting when cementation services will be required. On most land wells, a substantial part of the overall cementing charge will be standby time for the cementers. This must be kept to the minimum possible.

Electric logging $30 000 4.9 per cent

In most companies, Petroleum Engineers will stipulate the electric logs that they require for a given well. The cost control on this is usually down to the tendering exercise and supply and demand. The Drilling Manager can try to ensure that no spurious logs are being run, but in most cases, he will be told what to run and have little or no input into the logging suite design.

Well engineering $50 000 8.1 per cent

The cost of engineering a well depends, basically, on whether it is done in-house or by using a specialist company. For operators with an intermittent drilling programme, the latter is probably more cost-effective, however, continuous drilling programme should probably be programmed in-house. When done in-house, the cost of well engineering will further depend on the manpower allocation, their costing, their effectiveness and the systems in place to allow quick and effective well programming. This cost centre, therefore, is under the direction of the Drilling Manager.

Overheads $70 000 11.4 per cent

Overheads are under the direction of the drilling manager, to some extent, but are also influenced by the assignment of Petroleum Engineers, Geologists and Administration staff within the operator's organisation. The Drilling Manager can ensure that he has no excess staff assigned to the well but beyond this, has little influence.

2.3 Comparison of cost breakdown between onshore and offshore wells

Comparing both sets of cost breakdown, we can see the following differences:

1. On the offshore well, the fixed costs (38 per cent) are less than the daily costs (58 per cent).
2. On the onshore well, the fixed costs (57 per cent) are more than the daily costs (41 per cent).

This is a fairly typical situation and its basic implication is that the economic influence of the Drilling Supervisor's skills in drilling a quick well are felt more on offshore than onshore. This is not to say that he does not have a major role in controlling well costs on land wells, but serves to indicate just how important it is to have the right office team of Drilling Manager and Drilling Engineer, if these costs are to be controlled effectively.

Offshore wells generally cost a lot more than land wells, and the example shown indicates that to save a day on the well would reduce the well costs by $50 000. A day saved on the shallow land well studied only represents a $12 500 saving. The difference between these two figures indicates that, when looking at time-saving services and equipment to be used on the rig, we have to cost out the time saved against the cost of services, rentals or purchases. Consequently, some services and tools which make great savings offshore might never pay off on land.

3. AUTHORISATION FOR EXPENDITURE (AFE)

The AFE is the tool that is used for predicting the cost of a proposed well. The accuracy of the AFE depends on the amount of available information used to construct it. As operators, we need to know how much a well is going to cost if it is dry, tested or completed. Consequently, AFEs should be broken down into sections to allow us to see at a glance how the various well options compare financially. In this section, we look at the component parts of AFEs for onshore and offshore wells. At Norwell, we have a standardised 100-point AFE for both onshore and offshore wells. This is probably overkill as in most wells there are probably only 60 or so cost centres, but having the additional codes for special operations built into every well AFE makes it easier to carry out post-well assessment and cost comparisons between wells.

3.1 AFE components

Both the onshore and offshore AFEs are broken down into the following sections:

preparation
drilling and abandonment
testing
completion

3.1.1 Preparation

This part of the AFE covers the costs incurred to the point at which the rig is brought on to location. For onshore wells this would include site building and well engineering as the main cost centres. For offshore wells, the main cost centres are site surveying and well engineering. Included in this section should also be all the costs required to bring the location back to its original condition.

3.1.2 Drilling and abandonment

This is the 'dry hole' drilling component of the well. It assumes drilling to TD, logging and finding nothing of interest. The well is, therefore, proposed for abandonment and costed accordingly.

3.1.3 Testing

This is the additional cost incurred by a testing programme. It is not merely the testing cost charged by the testing company but must also include all the ongoing daily costs associated with the rig such as:

rig day rate
fuel oil
site personnel
office personnel
office overheads

3.1.4 Completion

This is the further additional cost incurred following testing once the decision to complete the well has been made. As with testing, the cost centres are not only the cost of completion equipment and services but also the costs of:

rig day rate
fuel oil
extra casing string if run
perforation
site personnel
office personnel
office overheads

3.2 Estimating costs

If there are similar, recent wells in the area to be drilled, most costs can be estimated fairly readily. If, on the other hand, you are planning a well in a new area, then the task is much harder. By calling up the following service companies and asking for budgetary figures, the main cost centres can be addressed:

drilling contractors
mud loggers
electric logging companies
mud companies
cementing companies

bit companies
casing companies
wellhead companies
tool rental companies
coring companies

Most service companies will be pleased to provide figures for AFE budgets and talking to them serves a secondary purpose of updating your knowledge of the demand for certain services and any new deals or equipment that is available. It also allows the service company to express an interest in the work which will be put out to tender at a later date.

The Time Depth Graph created for the Drilling Programme provides an estimate of the days to be spent on the well. By costing in the charges for these days, the AFE begins to take form. It is difficult to fix charges such as coring on an exploration well with the limited knowledge available regarding formations to be drilled so some assumptions must be made. The AFE could either include one 20 m core or several runs.

Similarly for the testing programme and completion programme some assumptions must be made. It is, therefore, good practice to list the assumptions which have been made as a postscript to the AFE. For example, AFE assumes four days open hole testing, or AFE assumes single 3½ inch H_2S tubing completion with single permanent packer. After estimating costs, a contingency factor should be built into the AFE. This can be in the form of a lump sum or as a percentage of well costs.

3.3 AFE examples

The following are examples of the AFEs used in Norwell's operations and have proved to be generally effective. Most working AFEs will be of a similar format.

LAND WELL
Initial AFE (for budgetary purposes)

Cost code nos	Code description	Preparation	Drilling and abandonment	Testing	Completion
01	Site survey				
02	Site rental				
03	Site purchase				
04	Site preparation				
05	Site prep. management				
06	Additional Porta-cabins				
07	Additional tankage				
08	Conductor driving				
09	Communications				
10	Rig mobilisation				
11	Rig day rate				
12	Rig demobilisation				
13	Public meetings				

Cost code nos	Code description	Preparation	Drilling and abandonment	Testing	Completion
14	Wellhead				
15					
16					
17	20″ conductor				
18	13⅜″ casing				
19	9⅝″ casing				
20	7″ casing				
21	Liner				
22					
23					
24	13⅜″ casing accessories				
25	9⅝″ casing accessories				
26	7″ casing accessories				
27	Liner hanger				
28	Liner accessories				
29	17½″ rock bits				
30	17½″ PDC/diamond bits				
31	16″ rock bits				
32	16″ PDC/diamond				
33	12¼ rock bits				
34	12¼ PDC/diamond bits				
35	8½″ rock bits				
36	6″ rock bits				
37	6″ PDC/diamond bits				
38	Bit accessories				
39					
40	Drilling consumables				
41	Lubricants				
42	Fuel oil				
43	Drilling equipment rental				
44	Mud equipment rentals				
45	Equipment transportation				
46	Solid waste disposal				
47	Liquid waste disposal				
48	Misc. consumables				
49					
50	Mud logging				
51	Specialised equip. running				
52	Mud engineering				
53	Mud chemicals				

Cost code nos	Code description	Preparation	Drilling and abandonment	Testing	Completion
54	Cementing service				
55	Cement				
56	Cementing additives				
57	Casing running				
58	Directional Drilling				
59	MWD				
60	Well surveying				
61	Mud motor and turbine				
62	Fishing				
63					
64	Coring				
65	Coreheads				
66	Core analysis				
67	Wireline logging				
68	Reservoir sample analysis				
69	Velocity surveying				
70					
71	Downhole production testing				
72	Surface production testing				
73	Electronic gauges				
74	Perforating				
75	Prod. test consumables				
76	Production test rentals				
77	Acidisation/fracturing				
78	Gravel packing				
79	Coil tubing				
80	Nitrogen services				
81					
82	Christmas tree				
83	Completion tubulars				
84	Completion tubing acc.				
85	Completion tubing running				
86	Completion fluids				
87	Completion wireline ops.				
88	Completion perforating				
89	Completion consumables				
90	Completion rentals				
91					
92	Suspension				

Cost code nos	Code description	Preparation	Drilling and abandonment	Testing	Completion
93	Abandonment				
94	Site clean-up				
95	Site accommodation & travel				
96	Materials base				
97	Site personnel				
98	Office personnel				
99	Office overheads				
100	Insurance				
	Sub totals				

OFFSHORE WELL
Initial AFE (for budgetary purposes)

Cost code nos	Code description	Preparation	Drilling and abandonment	Testing	Completion
01	Rig rate				
02	Site survey				
03	Tow and anchor handling				
04	Mobilisation				
05	Anchors				
06	Anchor chain				
07	Buoys				
08	Rig location survey				
09					
10	Guide base				
11	Conductor wellhead housing				
12	20″ wellhead equip.				
13	13⅜″ wellhead equip.				
14	9⅝″ wellhead equip.				
15	7″ wellhead equip.				
16					
17	30″ conductor				
18	24½″/20″ casing				
19	18⅝″/16″ casing				
20	13⅜″ casing				

Cost code nos	Code description	Preparation	Drilling and abandonment	Testing	Completion
21	9⅝″ casing				
22	Liner				
23					
24	24½″/20″ casing accessories				
25	18⅝″/16″ casing accessories				
26	13⅜″ casing accessories				
27	9⅝″ casing accessories				
28	Liner accessories				
29					
30	Rock bits				
31	Strapax bits				
32	Drilling consumables				
33	Lubricants				
34	Fuel oil				
35					
36	Eqiup. rentals (drilling)				
37	Equip. rentals (mud)				
38	Supply boat				
39	Standby boat				
40	Telecoms				
41	Shore-side medical backup				
42	Weather forecasting				
43	Helicopter fuel				
44	Offshore catering				
45	Waste disposal				
46	Inspection services				
47					
48	Underwater TV				
49	Diving/ROV				
50	Mud logging				
51	Specialised equip. running				
52	Mud engineering				
53	Mud chemicals				
54	Cementing service				
55	Cement				
56	Cementing additives				
57	Casing running				
58	Directional drilling				
59	MWD				

Cost code nos	Code description	Preparation	Drilling and abandonment	Testing	Completion
60	Well surveying				
61	Mud motor and turbine				
62	Fishing				
63					
64	Coring				
65	Coreheads				
66	Core analysis				
67	Wireline logging				
68	Reservoir sample analysis				
69	Velocity surveying				
70					
71	Downhole production testing				
72	Surface production testing				
73	Electronic gauges				
74	Perforating				
75	Prod. test consumables				
76	Production test rentals				
77	Acidisation/fracturing				
78	Gravel packing				
79	Coil tubing				
80	Nitrogen services				
81					
82	Christmas tree				
83	Completion tubulars				
84	Completion tubing acc.				
85	Completion tubing running				
86	Completion fluids				
87	Completion wireline ops.				
88	Completion perforating				
89	Completion consumables				
90	Completion rentals				
91					
92	Suspension				
93	Abandonment				
94	Site clean-up				
95	Demobilisation				
96	Materials base				
97	Offshore personnel				

Cost code nos	Code description	Preparation	Drilling and abandonment	Testing	Completion
98	Office personnel				
99	Office overheads				
100	Insurance				
	Sub totals				

4. COST CONTROL DURING DRILLING

It is vital that the operator and partners be kept informed of the day-to-day running costs of the well. Consequently, most daily drilling reports will include a cost element. Once the AFE has been drawn up, it will be apparent how many working codes within the AFE are being used for the well. Some codes will have no cost allocation since they do not apply to the well. By listing only the working codes on a single sheet, a daily cost sheet can be constructed which can be filled in by the Drilling Supervisor or Site Drilling Engineer. As with the AFE, it is sometimes difficult to put precise costs on certain tools or operations. However, the best estimate possible must be made on a daily basis if costs are to be controlled. The cost of phones, faxes and telexes will not be known until the bill comes in. However looking at earlier wells can give a rough guide; if in doubt, allow a little extra.

Chapter 6

DRILLING CONTRACTS AND TENDERING

1. CONTRACT TYPES

Drilling contracts generally fall under four main headings:

1. Day rate contracts
2. Modified day rate (footage bonus) contracts
3. Footage contracts
4. Turnkey contracts

1.1 Day rate contracts

This is the most common type of contract used worldwide. The drilling contractor is paid a specified sum by the operator for each day that he spends on the well. There is no performance bonus on this type of contract but the contractor can be penalised for negligence.

Most drilling contractors like this type of contract as there is little downside for them. They supply the equipment and personnel and, providing that they do not make too many mistakes, they can enjoy a regular income for as long as the well takes. There is no financial incentive for the contractor to drill more quickly or handle BOPs more efficiently; in fact the converse is true.

When operating this type of contract, the operator's Drilling Supervisor on site must be especially vigilant to detect delaying tactics on the part of the contractor. In large areas, one company rig working alongside several contractor rigs can act as a lateral form of control on drilling performance. The company rig is operated by staff who get paid by the month and not by the contract. They have no incentive to slow things down. They drill ahead as best they can and usually set the standards that the contractors must endeavour to match. This does not happen in every case, but usually the company rig is a useful yardstick by which to check contractor's performance.

Day rates are usually broken down into four groups:

1. Operating
2. Reduced
3. Special
4. Zero

Operating rate is normally applied to rig utilisation. It implies that the contractor's equipment and personnel are fully utilised. This rate covers activities such as drilling, tripping and casing runs.

Reduced rate is used when the contractor's equipment and personnel are not being fully utilised. During electric logging, for example, most of the contractor's equipment is shut down therefore the reduced rate will apply. Other operations which can fall in this category include WOW rig moves and ballasting operations. This rate is usually a few per cent cheaper than the operating rate.

Special rate comes into force in many circumstances. A typical example would be when the allowable repairs downtime allowance was just exceeded in a given month.

Zero rate is when no payment is made to the drilling contractor. The usual reason behind this is negligence by the contractor causing operational delays.

1.2 Modified day rate (footage bonus) contracts

The purpose of this contract type is to encourage the drilling contractor to reach TD more quickly than he might on a conventional day rate. The operator will retain conventional control over this type of contract, however there will be a bonus for the contractor if he reaches a certain depth within an agreed time scale. This type of contract can be attractive to both the operator and drilling contractor but is not applied very often in practice.

1.3 Footage contracts

These contracts usually are given in reasonably well-known areas. A specified rate per metre or per foot drilled is negotiated for a well of a certain depth. With this contract style the drilling contractor has a direct incentive to drill the well faster.

When offering a drilling contractor a footage contract the operator should clearly list any special terms with which the contractor must to comply for the specific well. An example would be the stipulating of a maximum tripping speed to prevent pressure surges, were it known that this was detrimental to the likely productivity of the reservoir.

It is usual to keep a company drilling supervisor on the rig at all times during footage contracts, although many of the traditionally operator decisions such as bit selection and drilling parameters will be made by the drilling contractor.

1.4 Turnkey contracts

With this kind of contract the operator pays the drilling contractor a lump sum to drill a well of a certain depth in a given area. It is incumbent upon the drilling contractor to procure all the well head and casing requirements, organise the third party services and generally fulfil all the normal operator's roles on the well.

It is probable that the operator will insist on rights of inspection at any time on the well. However, the operator will have no input into the day-to-day operations **unless** he has clearly stated in the turnkey contract that he wishes to retain that right. Since, in most areas of the world the operator can never give away the responsibility for oil spills in his license block, turnkey drilling has had a limited impact on the market. Notable exceptions to this are USA and China where local legislation makes turnkey drilling attractive.

The usual criticism of turnkey drilling by operators is their lost of control over operations. This is true with most existing turnkey contracts, but as mentioned earlier in this section, it is the operator who writes the contract. If you want control, write it into the contract. Most oil companies have very definite standard procedures covering BOP, stack testing and drilling practices. It is possible to write a turnkey contract insisting on these and ensuring that they are monitored. If the contractor did not like the contract then he would not sign it so nothing would be lost.

The usual criticism of turnkey drilling by contractors is their exposure to the risks of downtime or well problems. Consequently, they might insist on 'hole problem' or 'fishing' escape clauses from the turnkey contract during which they go on to a more conventional day rate. Also most contractors do not actively collate data bases on drilling information and are not involved in drilling optimisation on a regular basis. It is therefore difficult for them to adopt the conventional operator's role immediately, which is required for turnkey drilling.

As the technical competence of drilling contractors improves, turnkey drilling in some form will probably increase in popularity but with more detailed specific contracts than those of today.

1.5 Summary

The operator should select the contract style which suits his purpose best. The design of the drilling contract should be under continuous review to take advantage of changing conditions. In a large development with several strings running on a long term basis it is desirable to apply different contract styles in order to determine which is most effective for prevailing conditions.

When a drilling contractor is put on a footage contract, he invariably goes faster. This is an interesting situation since with the operator having less input into the drilling parameters, work progresses more quickly. Since the operator's own rigs probably out-perform the contractor's rigs, this in turn implies that the conventional drilling contract promotes inefficiency. By allowing the drilling contractor to perform periodically on an incentive or footage basis, a comparison between this performance and normal performance can be made.

If BOP handling takes 50 per cent less time on an incentive contract then one wonders why it took so long on the conventional contract. Allowing the contractor to drill occasional incentive contract wells gives the operator leverage when negotiating subsequent conventional contract wells.

The safety implications of incentive drilling must always be considered. Does the contractor speed up due to increased efficiency or by taking short cuts? Not all short cuts are bad but some do endanger personnel.

2. CONTRACT FORMAT AND MANAGEMENT

The contractor rig contract on most wells represents 30–40 per cent of the overall well costs. To ensure that the operator's interests are best protected, the drilling contract is drawn up by the operator and issued complete to drilling contractors. Drilling contractors submit their bids for the work, based on the issued contract. Upon signing by both partners, the contract then forms the basis of the working relationship between operator and drilling contractor. Both partners are bound by the contract, therefore it is vital that the design of the contract is such that there are no loopholes or deficiencies which could cause the operator to be exposed unnecessarily during the execution of the contract.

Note: The operator writes the contract, so if he has cause to criticise the contract during its execution then he only has himself to blame!

The purpose of the following text is to ensure that this never happens. The contract must cover all aspects of the operator/contractor relationship. In a well-written contract there are no 'grey areas', everything should be 'black and white'. Contract formats vary from area to area to suit local conditions. However, certain clauses are common to all drilling contracts.

Examples of these clauses or groupings are:

1. Agreement
2. Personnel, equipment, services and supplies
3. Compensation
4. Manner of compensation
5. Quality control or performance
6. Liabilities
7. Drilling methods and practices
8. Insurance
9. Special conditions
10. Confidentiality
11. Patents
12. Assignment of contract
13. Laws, rules and regulations
14. Arbitration
15. Signatures

1. Agreement

The agreement should describe the objective of the contract. It should clearly state when the contract will begin, its duration and termination date.

2. Personnel, equipment, services and supplies

This clause lays down what the contractor is expected to supply. Most contracts in this clause refer to appendices which are included with the contract, where a list of equipment or personnel is offered by the contractor and included in the tender document.

The operator should clearly state what personnel and equipment it will provide. A format should be established to cover any additional equipment which might be required. Usually this amounts to allowing the contractor to add on a handling charge if it purchases additional equipment to fulfil the purposes of the contract.

3. Compensation

This details how much money the contractor will receive for this services. All aspects of the operation must be covered in this section and it is better to apply too much rather than too little detail to describe the grouping into which services will come. A typical breakdown would be:

Mobilisation charges:
Rates:
- Operating:
- Repair:
- Inspection:
- Maintenance:
- Negligence:
- Force majeure:

Demobilisation charges:

Also included in this section will be the contractor's liability to taxation and any other charges. If the rates are adjustable then any adjustments should be explained. Some rates, for example, increase annually according to a cost of living package. In this case, the cost of living package must be clearly defined.

4. Manner of compensation

This clause explains how invoicing must take place and the delay after which payment will be made.

5. Quality control of performance

In this section the operator reserves the right to insist that the contractor's equipment and personnel can fulfil the commitment made in the tender document. In practice, this means providing access to the hardware and to records. For a semi-submersible rig, structural surveys, inclining test records and certification of drilling staff would have to be checked. All tubular goods and lifting equipment certificates and planned maintenance records would have to be available for inspection. Usually, a rig audit team will visit the rig prior to awarding the contract to inspect all of the above.

Most operators will include in this section an opportunity for the operator to take over the operation and contractor's personnel for reasons which will be defined. The operator will normally reserve the right to insist that the drilling contractor removes undesirable crew members. The method by which the operator wishes to effect this must be clearly defined.

6. Liabilities

This section defines the operator's and the contractor's liabilities. The bulk of this section will represent the contractor holding the operator harmless in respect of damage to his equipment and vice versa. There may be details of how the operator's equipment must be serviced under the contract by the contractor.

To cover the cost of any contractor's equipment lost downhole, the contract will define the formula by which the present value of the equipment is determined. Generally speaking, the contractor's liabilities tend to be limited in most cases, whilst those of the operator are considerable. An average pollution liability would be limited to $100,000 for the contractor, whilst the operator's liability to the licensing authority or government would be unlimited.

7. Drilling methods and practices

This defines the safe working practices that are expected from the contractor. Details such as special BOP testing, tripping, mud monitoring and reporting should be clearly defined.

8. Insurance

This establishes the level of insurance cover that the operator and contractor will have to carry.

9. Special conditions

Any possible unexpected circumstances should be clearly defined in this section. The action and responsibilities must be made clear in each case. *Force majeure* will come under this section and will cover all eventualities which are beyond the control of the operator or contractor.

10. Confidentiality

The contract must clearly lay down what the contractor may disclose about the well it is drilling.

11. Patents

This section indemnifies each party from patent infringement by the other party for the duration of the contract.

12. Assignment of contract

This section can give the operator the right to assign the contract to another operator (Farm out agreement) or can prevent this from happening.

13. Laws, rules and regulations

This defines under which government laws the contract will operate. Usually, this will be the area where the drilling is taking place. However, in frontier areas another country's laws might be used. Typically, this would be the home country of the operator, for example, AGIP – Italy, Shell – Holland, ESSO – United States.

14. Arbitration

In the event of dispute of the contract, the arbitration process should be clearly defined in this section.

15. Signatures

This section has space for signatures and witnesses for the operator and contractor. Note that each page of the contract should be initialled by both parties as well as any changes to the contract.

3. CONTRACT NEGOTIATION

Negotiation techniques are the same whether you are renting a rig or buying fruit in an Indian market. To negotiate successfully, you must know exactly what your own resources are and what you need. Furthermore, you must also try to know as much about the contractor as you possibly can. This section looks at getting the best deal from a contract by both sides. To do this, it is important to consider what both sides want from the contract.

Items	Operator	Contractor
Rates	cheap as possible	expensive as possible
Rig choice	highest rated rig	lowest rated rig
Service	maximum service	minimum service
General	quiet life	quiet life

Since only one item, quiet life, is desired by both sides, it can be dealt with first and dispensed with.

To gain a quiet life, the contract should be clear and uncomplicated. It is in neither party's interest to have a contract which needs a team of QA/QC men, time and motion study men, international financiers and accountants to work out the rate applicable at the end of each month for invoicing. Contractual disputes take up time and money, so both sides are pleased to have a simple contract format.

Drilling contractors like oil companies are in business for one thing only; to make money. Oil companies have several ways of making money, but drilling contractors have only one. The contract, therefore, is vital to the contractor's profitability and hence even a mediocre contractor will be very aware of the methods of extracting funds from a contract.

3.1 Contractor's expenditure

Before looking at how the contractor can make money from a contract, his expenses should be understood. These costs can be broken down into 11 groups.

1. Interest payments on rig
2. Insurance cost of rig
3. Repair and maintenance
4. Catering
5. Personnel
6. Training
7. Administration
8. Marketing
9. Capital replacement
10. Inventory
11. Contingency

If the contractor can, by manipulating the contract, get the operator to assist in any of the above expenses, then the contractor's profitability will increase. This concept must be fully understood by the operator to ensure that he is aware that what seems a small concession by the operator, could increase the contractor's profitability considerably. When bargaining you must know what your 'chips' are worth.

1. Interest payment on rig

Most drilling contractors borrow money from banks to buy rigs. The banks will set an interest rate for repayment. The drilling contract has no influence over this and therefore this cost centre can not be affected by contract manipulation.

2. Insurance cost of rig

The rig owners must insure their rig. The rates of insurance vary very little from block to block in a given area, therefore there is no room for manoeuvring in this cost centre.

3. Repair and maintenance

The drilling contractor bears all of these costs. If he can reduce them then his profitability will increase. Any clause or condition that the contractor can insert into the drilling contract which allows him to spend less money on repair and maintenance will increase his profitability.

Clauses which appear regularly in contracts such as '30 minutes per day oiling and greasing time' are allowing the contractor a cost-free facility to repair equipment on the operator's time. How many operator owned rigs need 30 minutes per day oiling and greasing? The drilling operation is by its very nature intermittent, hence allowing periodic lubrication daily.

Other clauses, which greatly increase the contractor's profitability, are those where the operator pays for certain spare parts. An example of this would be if the operator paid for all rubber goods in the BOP stack. This action would be taken by the operator in an attempt to ensure no scrimping on the quality of BOP equipment. An unscrupulous contractor can, however, keep his entire fleet of rigs supplied with BOP rubber goods from this one contract. This can be prevented in several ways. Auditing rubber goods consumption over a period of years or comparing it with the operator's own rigs can quickly highlight any misuse of the contract intention. Another way would be to put a cash limit on purchasing in a given period. However, this in itself could be counter productive to the contract intention. Contrast clauses should never be open-ended, since the temptation can prove too great for some to resist exploiting them.

It is much neater to built all these miscellaneous charges into the basic day rate and have severe penalty clases (such as zero rate) as soon as shoddy maintenance standards make themselves felt.

4. Catering

Most contracts include catering for the contractor's crews. Contractors can increase their profitability by reducing catering costs. An army marches on its stomach and drill crews are not a lot different.

Within the contract, it must be clearly shown what the standards of catering will be. If a third party caterer is being used, the payment details should be included to illustrate what the contractor's intended expenditure will be. There must be a mechanism whereby the operator can insist on a good standard of catering from the contractor at zero additional cost to the operator.

Contractors are allowed to charge catering costs to the operator for the operator's staff over and above a stipulated number (normally six or so). This figure can be a profit centre for contractors. During production testing, for example, the operator could have 20 or 30 extra staff on board and a nightly charge would be made by the contractor to accommodate them. A few extra dollars per head which slipped by at the contract reading stage could add up to a considerable sum by the end of the year.

5. Personnel

To operate the rig, the contractor needs onshore and offshore personnel. Good staff do not come cheaply and the personnel costs represent a major proportion of the contractor's expenditure. His costs can be reduced by two means: employing fewer people, or employing the same number of people but paying them less.

It is therefore in the contractor's interest to bid for a contract with minimum personnel, in the hope that additional personnel can be charged for outside the basic rig contract. It is furthermore not in the contractor's interest to disclose the rates that he is paying on or offshore staff, or to be held to certain levels. It is also not in the contractor's interest to be held to using his own staff rather than independent labour pool staff.

6. Training

It is usually incumbent upon the contractor to ensure that his staff have the appropriate certification for their positions. Unless the drilling contract stipulates differently, then that is where the training effort might stop. Clearly, training is an expense for the contractor and if he can get out of spending any extra money on it he can improve his profitability. If the contractor can get the oil company to train his staff at the oil company's expense then this again improves the contractor's profitability.

7. Administration

The cost of administering a drilling rig is considerable. To the operator, a smooth contractor administration is required to support an efficient rig. However, any excess administration will be paid for in the day rate directly or indirectly by the rig crews receiving less of the salaries pot.

It was noticeable that as the 'cold wind' blew through the drilling industry in 1986/87, many administrations survived, whilst the offshore crews bore the brunt of the economic cuts. Most contractors would like to disguise the cost of administration by quoting a total salary cost but it is not necessarily in the operator's interest to allow them to do so.

8. Marketing

These costs cover publicity, publications, services and the salesmen that contractors use to market their rigs. These costs can be quite high and consequently longer term contracts are attractive to contractors as they reduce the marketing effort and therefore cost. It must be realised that as with all costs incurred by the contractor, the operator pays for them in the end.

9. Capital replacement

Rigs tend to be written down over the years and, if the drilling contractor intends to stay in the market, he must plan to purchase new rigs ultimately to replace the old ones. To purchase rigs, the contractor needs capital and capital is accumulated over the years from other rig contracts.

10. Inventory

All spare parts that a contractor has sitting on the shelf, either on the rig, or at his yard, cost him money. Clearly, carrying a $0.5 m inventory on a rig is cheaper to a contractor then carrying a $1 m inventory. The skill in the equation is ensuring that sufficient parts are carried to cope with servicing and breakdowns, whilst still maintaining a minimum stock level. As far as possible, the contractor will try to minimise the inventory that he has to carry.

11. Contingency

Contingency costs cover such items as unexpected major equipment failure or redundancy payments for crews. Any concession that the operator makes, which reduce contingency costs, adds to the potential profitability of the contract for the contractor.

3.2 Contractor's income

In every aspect of the contract that the oil company pays out, the contractor **can** make a profit. Most rig comparisons refer only to the day rate, as if it were the only payment involved. In fact, many payments can be paid by the operator and each of these must be finely tuned to ensure that no exploitation is taking place. Payment can be divided up into 12 forms, as follows:

1. Mobilisation fee
2. Day rate
3. Charge for additional personnel
4. Charge for additional equipment
5. Built in repair time
6. Handling charges
7. Escalation charges
8. Charges for replacing damaged equipment
9. Charges for consumables
10. Charges for accommodating additional OPCO personnel
11. Charges for OPCO modifications
12. Demobilisation fee

As can be seen from the above list, whilst the day rate is the major cost centre, the drilling contractor still has a lot of ways of extracting additional income from the unwary operator.

Now we will look in more detail at the sources of income that a drilling contractor enjoys.

1. Mobilisation fee 12. Demobilisation fee

What exactly are we paying for? In practice, it is a fee to cover the extra charge facing the contractor to move a rig or start up or lay off a rig. The contractor faces real costs for this, so a payment of some sort is required, either directly as mob/demob payment, or indirectly as an additional day rate payment.

On land wells, the mobilisation fee covers the time until the rig is ready to spud. On offshore wells, this is not always the case. Semi-contracts, for example, will often pay a mobilisation fee and then standby rate until the rig is towed to location and moored up.

During certain periods, bad weather can be expected (October is usually very bad in the North Sea) and anchor handling problems can be expected. It is up to the operator to decide how much responsibility he wishes to take for the downtime so incurred. The most effective means of limiting this responsibility is to pay a mobilisation fee which takes the rig from its original location until the rig is ballasted down and moored at its new location. Any delays are then to the contractor's cost.

The operator must protect his interests and try to fix costs at every stage possible to maintain tight fiscal control on the well.

2. Day rate

In life you only get what you pay for. The day rate is unavoidable, however, we must determine what each rig's day rate involves, to establish value for money. If the well only requires 10 m stacks then 10 m rigs will offer much better value on the day rate than 15 m rigs. A 15 m system costs much more to buy and maintain, therefore will cost more on a day rate than an equivalent 10 m system. If both types of rig are being offered at the same price then, either one is charging too much, or the other is charging too little and scrimping on other features.

Included in the day rate is the repair and maintenance and inventory. Documentation supporting the systems in place must be studied to determine the professionalism of the contractor when comparing options.

3. Charge of additional personnel

The basic contract should include all the personnel for the normal work load. Additional personnel should only be paid for at market rates. It does not take long to determine what market rates are and this must be done. The contract must stipulate the procedure whereby additional personnel can be requested and supplied and the terms under which this will happen.

4. Charge of additional equipment

As far as possible, this cost should be avoided by building into the equipment list all equipment needed for normal operations. It is much easier to compare bids if all contractors are quoting exactly the same equipment list. Again, only market rates should be payable for additional equipment and a formula established to ensure that this will be done in practice.

5. Built in repair time

Putting the rig on standby rate for excessive downtime is no threat to a drilling contractor. In practice, the standby rates are only a few per cent cheaper than the operating rate. It is much more effective to have a contract whereby the contractor is allowed a few hours of repair at operating rate, after which the rig goes on zero rate. Clauses such as this allow the rig staff to get the support that they require from their base management.

6. Handling charges

If the operator asks the contractor to arrange or purchase something, then it is only fair that the contractor makes a handling charge. Usually, this is in the form of a mark-up, typically 5–10 per cent. In practice, sometimes there is a scale of charge and this represents the best approach. A suitable scale charge would be:

up to £100 000/month – 10 per cent, or
anything over £100 000/month – no extra charge.

In this way, the contractor receives a fair handling charge, although the operator's ultimate liability is restricted and controlled.

7. Escalation charges

These are best avoided, if possible, as they are fairly complicated to administer in practice. Typically, they are related to the RPI or a projected inflation rate. A flat rate is a much better deal all round and does not keep the accountants up all night.

8. Charges for replacing damaged equipment

If the operator's programme has damaged some of the contractor's equipment then it is only right that the operator should compensate the contractor. The operator must protect himself from sharp practices here and he can do so by laying down clearly in the contract, the terms under which payment will be made. For charges for replacing damaged equipment the following are required:

1. Invoice showing purchase date and price.
2. Maintenance records.
3. Inspection records.

Failure to supply these will mean that no payment will be made.

Payment will be made as follows depending on age of equipment:

0–6 months	new price
6–12 months	80 per cent of new price
1–2 years	50 per cent of new price
2–3 years	25 per cent of new price
thereafter	zero payment

The damaged equipment will become the property of the operator after payment is made. Clauses such as the foregoing make it clear to both sides what charges are allowed. This means that the contract is easy to administer.

9. Charge for consumables

It is better to build all consumables into the normal day rate as they tend to be open-ended charges.

10. Charges for accommodating additional OPCO personnel

These charges should not be inflated and as before, the market rates should be established and laid out in the contract draft.

11. Charges for OPCO modifications

If the operator wishes to modify the contractor's rig or equipment then he must pay for it. In the contract, the terms under which modifications could be carried out must be clearly stated. Two outside engineering companies should be approved by both sides, as alternatives to the drilling contractor's in-house expertise, to provide services for OPCO modification. If only the drilling contractor is allowed to modify his rig then the operator faces a monopoly and has no bargaining power at all. If the work goes out to tender then the operator can establish the market rate for the work and use this to control costs.

3.3 Negotiating contract terms

RULES FOR NEGOTIATION

1. **Know what you want**
2. **Know what the contractor wants**
3. **Know what you are prepared to pay**
4. **Know what the contractor is prepared to take**
5. **Play down the contractor's strengths**
6. **Accentuate the contractor's weaknesses**
7. **Keep calm at all times**
8. **Exploit but never humiliate the contractor**

In the previous sections we have looked at the ways in which a drilling contractor makes money and spends money. Knowing this information gives the operator insight into the contractor's affairs. Further specific information should be established prior to negotiating a contract:

1. What suitable rigs are available?
2. What alternative work is available for these rigs?
3. What are market forces?
4. What are individual contractor's priorities?

3.3.1 What suitable rigs are available?

This can be established by subscribing to specialist journals which list available rigs, or by consulting the contract drilling companies directly.

Quite often rigs are working for other operators on a fixed number of well plus option deal. Most contractors will bid these rigs as potentially available, but the option clause could prevent them from meeting your deadlines. A phonecall to the operator concerned could eliminate some of the doubt surrounding possible option jack-up. Clearly a rig which is definitely available is more attractive than one which is potentially available. A list should be made of the available rigs that could do the work required at the time required.

3.3.2 What alternative work is available for these rigs?

This is slightly more difficult than determining which rigs are available. A news collection service, either formalised, or informal, should be used to gather information into other operators' plans.

Service companies usually have a good idea about what is going on and can provide useful information about other operators' intentions. Talking to other operators at operators' meetings can provide more details of their plans.

To negotiate with contractors, you must know what alternatives are available to the contractor. If he has a lot of alternatives, then it will be more difficult to strike a bargain than if there is little work available.

3.3.3 What are market forces?

This is even more difficult to predict, but an attempt must be made to quantify the prevailing forces. Market forces can make themselves felt in many ways, some of which are outlined below.

Fashions change and rigs come into and or out of favour. There is a current trend in the North Sea towards using jack-ups, especially large ones in areas which were traditionally semi country. If there is a large demand for jack-ups, then their prices will go up. If you are drilling in an area which could use either type in a strengthening jack-up market, then a better deal could be struck for a semi than a normal market would permit.

There is a finite number of rigs worldwide. If a new area opens up to drilling, rigs must be brought from other areas to do the work. This can lead to shortage of rigs in the original area. This criterion can be thought of as a global extension of the alternative work scenario. It is necessary to know what is happening globally to negotiate successfully. If a new area is opening up in six months time, the contractor may favour a six-month contract at a bargain price.

Certainly the contractor tries at all times to predict the future markets as best he can. To move a rig from one area to another area costs money. There are towing charges and the rig earns no money during the move. Consequently, the contractor must see a greater potential revenue from the new area compared with the old area just to break even financially.

The easiest way to determine regional distribution is to subscribe to a dedicated specialist journal which keeps its readers up-to-date on developments. Third party service companies usually have good data bases on this information and should be consulted on an informal basis.

3.3.4 What are individual contractor's priorities?

To understand this, the motivation of the drilling contractor should be investigated. What is the contractor's intention? If he is in expansionist mode, then he may be very interested to cut his profit margins to secure regular work to fund his expansionist programme. If there is uncertainty about the future for the contractor, he may put his rigs on the market for sale. In this case, he will not be looking for longer term contracts.

Does the contractor wish to penetrate a new market? Does he, for example, have a reputation for running jack-ups but wish to get into semi drilling. In this case, he will probably subsidise the operation to get the first few jobs in the new market. To neglect the contractor's priorities is naive and expensive for the operator.

3.4 Summary

Answering the foregoing questions gives us an overview of the situation which it is vital to have prior to beginning negotiations. If at all possible, negotiation should be avoided, as it is time consuming and costly. We have seen in section 3. how the contractor makes and spends his money. To have influence in these matters, the contract that the operator issues must be specific and definite in order to reduce negotiated items to a minimum. To effect this we will look at the conventional day rate contract and how out theories can be applied in practice.

4. CONTRACT TENDERING

Having established our rig requirements for a given Drilling Programme, we then approach drilling contractors with this type of rig and ask if they wish to be included in the tendering exercise. From this information, a list of contractors to tender is drawn up and at this time a short-list can be made of the contractors that we feel can perform well.

Filling in tenders is a laborious and time-consuming exercise for the drilling contractor and also creates a lot of work for us in analysing the bids thoroughly. Consequently, the numbers of contractors invited to tender should be restricted to the five or six front runners.

4.1 Invitation to tender

Each tenderer should receive the following:

1. A letter of acknowledgement that they can return (by fax or telex) to confirm receipt of documentation and their intention to bid
2. A defined time scale in which to respond
3. A defined format in which to respond
4. A draft copy of the drilling contract that will be used, complete with a full rig and equipment specification requirement

They should be asked to provide the following:

1. A rig operations manual
2. Copies of all valid certification
3. Full equipment list
4. Maintenance programme
5. Rig inventory
6. Training programme
7. Safety programme and records
8. Résumés of staff

They should send their tenders, in triplicate, to the operator in a sealed envelope.

4.2 Response time

To be fair to the drilling contractor, approximately four weeks will be required for him to fill in the tender and provide the required documentation. One of the reasons for the delay is the fact that no two operators seem to use the same tender format. If a standard format was ever accepted by all operators, the tendering process would speed up to a week or two.

4.3 Receipt of tenders

Acknowledgement of receipt of tenders should be sent to the drilling contractors. Tenders should be kept sealed until the official hour of opening. Opening tenders should be a 'closed door' exercise. Upon opening the tender, the following analysis should be made:

1. Has the contractor provided all the requested information?
2. Has the contractor met the requested rig specifications?
3. Has the contractor met the requested rig equipment specifications?
4. Is the rig on offer currently working? If so, can spud date be guaranteed? If not, how ready is it for work?
5. What are the costs involved?

From this analysis, a short list can be drawn up. If costs and equipment are similar then the decision as to which to take will tend to be a subjective one. If a company has served you well in the past, then they will be in a favourable position. The converse is also true.

When the choice is down to one or two rigs, it is worth visiting each rig to perform a rig inspection. This will serve to confirm that all is as it has been bid and to get a general feeling for the rig. (chapter 10 covers rig inspection in detail.) Although computers are allowing us to create systems to cover most functions, when inspecting a rig, instinct is also very important. If a rig feels right it will probably work well.

Once a rig is selected, the contract terms and amendments suggested by the drilling contractor should be discussed and a mutually agreeable deal struck. At this point, the unsuccessful bidders should be notified and thanked for their submissions. All submissions should be filed and used as reference material for future rig requirements.

5. WORKING WITH DRILLING CONTRACTORS

To get the best value from a drilling contract, the operator's representative, who approves the contractor's invoices, cannot be too familiar with the contract. As mentioned earlier, the drilling contractor for all his faults, will know the contract in detail. Liberties may be taken if the operator's representative does not monitor the exact time breakdown of operations and its contractual implications.

On the rig, the operator's Drilling Supervisor should go through each day's IADC report and agree that the description and time breakdown therein is accurate, prior to signing it. If this is done, then the IADC report can be used for all costing purposes related to the drilling contract.

The drilling contractor will prepare an invoice monthly, based on the charges for that month. Typically, the charges will have been broken down into operating rate, reduced rate, etc. This invoice will be presented to the operator's Drilling Superintendent for approval. Prior to approving the invoice, the Drilling Superintendent must check the charges are valid for the month in question. To do this quickly, he needs to have kept on top of the daily time breakdown during the month and kept his own running totals.

Most contractual disputes and differences are over the interpretation of rate application and these must be resolved prior to approving the contractor's invoice.

Chapter 7

WELL PLANNING

The preparation of good Drilling Programmes is vital for safe and effective drilling operations. The planner must collate all available technology and apply it in the form of a programme, to meet the well objectives as safely, cheaply and efficiently as possible.

1. WELL PLANNING PROCESS

The well planning process can be broken down into stages. Each stage is discussed in more detail later on in this chapter. Chapters 13 and 14 show how the well planning process fits into the overall drilling operation.

1. Defining of well objectives
   Why is the well to be drilled? Will it be a wildcat exploration well or an appraisal or development well?
2. Obtaining consent to drill from authorities
3. Collection of data
   All relevant data from seismic surveys and offset well reports must be scrutinised to ensure that the optimum drilling programme is prepared.
4. Establishment of subsurface pressure regimes
   One of the main purposes of studying the collected data is to attempt to predict the formation pore pressures that are likely to be encountered and plan for them accordingly.
5. Establishment of fracture gradients
   Fracture gradients must be established in order to select casing seats.
6. Drilling programme preparation
   Using the collated data, the programme is constructed in such a manner that well objectives can be met.
7. Rig specification
   From the draft programme a specification of the rig required to drill the hole can be prepared. This is covered in detail in chapter 9.
8. Authorisation for expenditure (AFE) preparation
   Using the AFE well costs are budgeted. This is covered in detail in chapter 5.

1.1 Defining well objectives

Wells can be designated as one of the following three types:

1. Exploration wells
2. Appraisal wells
3. Development or production wells

Planning of the first two groups is initiated by exploration departments who will also define objectives. Drilling of development or production wells is usually initiated by production departments who will also define objectives.

1.1.1 Exploration wells

These are drilled so that information regarding the hydrocarbon bearing properties of formations in an area can be acquired. These are the most difficult wells to plan for since little information is usually available about their geology or the pressure regimes of the area. Consequently the well plan can only be prepared from seismic data and regional geology. Since this is the case, the casing must be designed so that it can take a contingency intermediate string. The wellhead should be able to accommodate this facility. In exploration wells, a comprehensive mud logging and data recording system is usually required, as is a full suite of wireline logs. The facility to DST or production test must also be accommodated.

1.1.2 Appraisal wells

These are usually drilled to determine the extent of geological structures and reservoirs following initial successful exploration wells. Appraisal well planning is simplified since offset well data should be available. Drill stem testing and production testing is likely. There is a possibility that the well could be used as a production well and if this is the case, the well design must accommodate this feature. Extensive electric logging requirements can be expected.

1.1.3 Development wells

These are the easiest types of well to plan for since a substantial amount of offset well data should be available. The casing programme should be designed for production tubing and completion equipment. Electric logging requirements should be minimal and in some cases, could be eliminated by using MWD techniques.

1.2 Obtaining consent to drill from authorities

The energy authorities in each country of operation award exploration licences to operators, whose responsibility it is to comply with all the local rules and regulations. The well programme must accommodate all these rules and regulations.

1.3 Collection of data

Most of the decisions during well planning are based on formation pore pressure. It is imperative, therefore, that pore pressures be known or estimated as accurately as possible. The two sources of well pressure data for well planning are geophysical/geological data and offset well data. Offset wells provide more accurate data, however, in the case of rank wildcats or wells in new areas, there will be no offset well data available. This means that pore pressure prediction will be done by analysing seismic data.

Seismic data is used in the exploration phase to locate potential reservoir traps and to estimate formation tops in the lithological column. It can be used to give a qualitative estimate of the formation pressure and hence an indication of any pore pressure abnormalities.

Offshore seismic data is also used to detect the possible presence of shallow gas bearing sands. In this case a high resolution seismic is performed, investigative down to a depth of some 1000 metres below the seabed.

Seismic data is acquired by creating acoustic waves using some form of explosion or implosion and measuring the time taken for the wave to travel down to a subsurface reflecting bed and then back to the surface. The point of origin of the wave on the surface is called a shot point and the returning waves are detected at the surface by a series of geophones or, if offshore, hydrophones, placed at known distances from the shot point. The velocity at which seismic or acoustic wave is propagated through a formation depends on the density and elasticity of the rock and the type of fluid occupying the pore space. The degree of compaction (i.e. relative depth) also determines the seismic velocity in a particular formation.

Prior knowledge of seismic velocities, in particular formations over a range of depths, can therefore enable fairly reliable predictions of formation lithology to be made. The presence of overpressured formations and an estimate of the magnitude of the overpressure can be predicted by studying the seismic velocity data from shale sequences. As the density and elasticity of shale increases with depth and compaction then the seismic velocity will increase. Overpressured shales are undercompacted, which, at a particular depth, will result in a lower density and elasticity than expected and therefore, a lower seismic velocity than expected. By analysing formation interval seismic velocities in normally pressured shales and comparing them with data from apparently overpressured shales at similar depths, the extent of overpressure can be estimated.

Shallow gas sands present one of the most difficult well control problems in offshore drilling operations. They tend to be high pressure and low volume and can unload a well very quickly with risk of toppling jack-ups, sinking floaters and causing fire and explosion.

In order to try and identify potential gas bearing zones at shallow depths detailed, high resolution seismic surveys are performed and analysed by what is known as the 'bright spot' technique. The data acquired over a survey grid with relatively small spacing between seismic lines, typically 150 metres, the grid covering an area of only a few square kilometres around the proposed well location and down to a maximum depth of investigation of about 1000 metres.

The fact that acoustic waves are severely impeded when travelling through gas charged formations produces strong seismic reflections. These show up as relatively high amplitude anomalies on the seismic section and are characterised by their visual appearance as bright spots. By mapping them on a horizontal section their spread can be studied and compared to the proposed well location. This high resolution seismic technique can detect gas charged sands as thin as 1 m.

The absence of bright spots does not necessarily mean there will be no shallow gas and conversely the presence of bright spots can be caused by an anomaly other than shallow gas. Even so, if shallow seismic surveys do indicate a bright spot at the intended well location, it would be prudent to change the location and avoid drilling into it.

It is not possible to predict formation pressures in shallow gas sands from the seismic data but it is quite possible, though not normally the case, that they are overpressured and should be treated accordingly. Shallow gas tends to be contained in low volume pockets and if allowed to flow freely through a well-designed diverter system, should quickly deplete and exhaust itself.

Structure maps are produced by the geologists from seismic and offset well data in order to evaluate the geology over large areas of interest. Interpretation of the seismic data can enable the geologist to identify subsurface structures that have the potential to trap hydrocarbon accumulations. Furthermore, by correlating formation tops from the lithological columns of offset wells and interpolating at the point of interest, a geological prognosis can be made of the proposed well.

If an offset well appears geologically similar to the well which is to be drilled, then preparing the Drilling Programme should be relatively straightforward with all the appropriate data to hand. In heavily faulted areas however, even wells drilled fairly close to each other can exhibit quite different problems and should be treated as wildcats.

1.4 Establishment of subsurface pressure regimes

In order to proceed with well planning, the subsurface pressure regime must be established. Most sediments will naturally have a pressure profile similar to that of seawater since they were laid down in a marine environment. These formations are said to have hydrostatic pressures. Higher pressures can be encountered for several reasons such as:

artesian wells
gas caps
salt beds
sediment compression
mineralisation
surcharged formations

formations can also be sub hydrostatic due to:

depletion
faulting
artesian wells

Formation pressures can be estimated from regional and local geology or by using production or outstep well data. If the well is a rank wild cat then an estimate of the pore pressure can be obtained from the seismic survey. Typically, geophysical interpretation will give a working pore pressure prediction that the Drilling Engineer can use for his calculations.

1.5 Establishment of formation fracture gradients

As with formation pressures, prediction can be based on anticipated geology and outstep well records. Most rocks of a certain type will exhibit typical characteristics. This can be used to assist in fracture gradient estimation. It is vital to have an accurate assessment of this so that the casing design can be done in an effective manner. Once a leak-off test has been carried out in the well, equations such as 'Daines' are used by employing values of Poissons Ratio for given formations to estimate probable fracture gradients at other depths in the well.

1.6 Drilling programme preparation

Drilling Programmes can be broken down into 11 main sections:

Section 1 Well details
2 Well objectives
3 Casing design
4 Wellhead selection
5 BOP requirements
6 Cementing programme
7 Deviation programme
8 Survey requirements
9 Mud programme
10 Bit and hydraulics programme
11 Evaluation requirements
12 Operational procedure and time depth graph construction
13 Site plan
14 Reporting requirements and contact numbers

All Drilling Programmes will contain the above information in some form. These sections are covered in more detail below. Specialised wells could also contain other relevant data.

SECTION 1: WELL DETAILS

This is a brief summary of the well location, type, depth, operatorship and ownership. A typical layout of this is shown below:

Well name	Godwana 3
Well type	Appraisal
Country	Moldavia
Block	20/12
Surface co-ordinates	

Subsurface co-ordinates

Target size	200 ft radius
Target depth TVDSS	10 000 ft
Target depth AHSS	10 000 ft
Water depth	200 ft
Operator	Alpha Oil Co (40%)
Partners interest	Beta Oil Co (20%)
	Gamma Oil Co (20%)
	Delta Oil Co (20%)
Name of rig	
Type of rig	15M jack-up
Seabed condition	Sand/silt flat
Expected spud date	3rd Quarter 1991

SECTION 2: WELL OBJECTIVES

Well objectives have been discussed earlier in this chapter. A typical format for setting out the objectives would be as follows:

> The Godwana 1 well is an appraisal well whose objective it is to establish the presence of the Huey, Louey and Duey sand stones at ± 10 000 ft.
>
> All three sand stones will be cored and depending on findings, be production and injectivity tested.
>
> After testing, the well will be either plugged back and abandoned or suspended for later use as an injection well.

SECTION 3: CASING DESIGN

Casing tubulars are put into a well bore for the following reasons:

1. To isolate troublesome or unstable formations which may include heaving shales, lost circulation zones and flowing halites.
2. To isolate different pressure or fluid regimes:
   (a) To protect fresh water horizons especially domestic water bearing sandstones in land wells.
   (b) To protect producing formations from mud and water contamination.
   (c) To protect weaker zones from breakdown caused by heavy muds whose hydrostatic is required for pressure control purposes in lower sections.
3. To control well pressure by containment of downhole pressure.
4. To provide a stable environment for packers, liner hangers etc.
5. To support the wellhead and BOP stack.
6. To confine produced fluid to the wellbore and provides a flow path for it.

A typical offshore well may have a 30 inch conductor, a 20 inch surface casing, 13⅜ inch and 9⅝ inch intermediate casing and a 7 inch production liner. Although the above is a common arrangement, there is a multitude of different combinations of casing sizes which the operator may choose to run if he desires and if the casing design allows it.

The following is a summary of the major functions for each type of casing:

Conductor seals off shallow unconsolidated formations and protects the wellbore from collapsing. It is the first string of casing upon which the BOPs can be mounted, usually being a diverter only, if it is preferred to handle gas at surface as opposed to venting at seabed. It must be able to support the weight of subsequent casing strings, the wellhead and BOPs. With the exception of floating vessels, the conductor may be pile driven or a hole drilled for it.

Surface casing will case off unconsolidated formations, and protect against shallow gas flows. Especially pertinent on land wells, this casing should ensure fresh water sands are not contaminated with drilling fluids or hydrocarbons from other zones. Near surface lost circulation zones can be cased off. Because the main BOP stack normally sits on this casing's wellhead, the installation of the surface casing gives full blowout protection for further drilling.

Intermediate casing will case off troublesome lost circulation zones, thus protecting them from the high pressures resulting from heavy muds or kicks when drilling deeper zones. They will nearly always be set in a pressure transition zone.

Production casing is the casing inside which the production tubing or testing string is run. It will separate and isolate the productive intervals from each other, and ensure influxes of undesired fluids do not occur.

Liners allow achievement of a production string at low cost because they hang off typically 150 m above the intermediate casing's shoe. They are often used as an additional contingency string, to be run if problems arise. Scab liners are run to protect casing from wear, or to repair damaged or worn casing. They are run on top of an existing liner. The well planner must ensure that the casing is designed accurately and sensibly, not only in the interests of well control, personnel safety, but also economy.

In addition to the above, the relevant government department may request seeing the casing design prior to granting approval for drilling the well. The operator may also have to justify the design to its partners.

Casing design incurs risk and the designer must reduce this to an acceptable level. Assumed loadings exerted on the tubulars will not be exact and the properties of the steel are variable, even though the material is within its designed specification. In addition to assuming worst case situations, the designer must incorporate design safety factors dependent on the reliability of his offset data.

At the end of this section, we discuss centralisation and casing running principles.

Exactly how much of this should be built into the Drilling Programme will be dependent on how comprehensively this is covered in the Drilling Operations Manual.

Pressure profile chart

A major section of the design work involves construction of the Pressure Profile Chart showing expected pore pressures and fracture gradients. It is the basis of the complete design and time should be taken to ensure that it is drawn correctly. Construction of the Pressure Profile Chart is especially useful for rough line sketching to look at the burst mode situation. Calculations can then be performed to verify the graphical solution and check the collapse and tensile conditions. The following minimum data is required:

1. The prognosed lithological column, including error bands on the tops of the formations.
2. Offset wells pore pressure data, including mud logging information, RFT and DST values and the pressures from any influxes. Details of mud weights used may prove useful.
3. Offset wells fracture gradient data, coming mainly from leak-off tests but also from lost circulation zones.

The pressures obtained from the offset wells must be depth matched to the relevant formation on the prognosed geological column, prior to plotting. In the event that no fracture gradient pressures are available, i.e. in rank wildcat wells, equations such as those used by Eaton can be utilised. In practice, this is rarely necessary.

The **overburden stress gradient** is a value quoted to indicate the pressure resulting from the compaction of strata above.

The Design Engineer will have to liaise with the Geologist to confirm that the wells he uses as major offsets are the most relevant to the planned well. Because of natural geology including faults, offsets which are closest may have very different lithologies or pressure regimes from the well being designed. Implications of using inaccurate offset well data may be that the casing is under designed to an unsafe level or over designed to the detriment of economy. A different offset may have to be used for the lower part of the well than what was used for the upper section.

Initial design approach

By the time the pore pressure chart has been constructed, the Drilling Engineer should have a reasonable idea of what casing sizes he will use and their setting depths. Choice of the casing OD will depend on availability of standard diameters, programmed hole sizes and likely rig choice. The final hole size for logging and the likely production string must both be given consideration. The Planner will have to liaise with Petroleum and Production Engineers at this early stage. For a normal offshore exploration well, it is recommended that a 8½ inch hole be the smallest diameter planned for because of drilling and evaluation difficulties encountered with a 6 inch hole. A 6 inch hole should only be planned for as a contingency.

The **setting depth** will depend on competent formations with high fracture gradients as indicated in offsets, lengths of open hole sections and requirements for cementing programmes.

A sense of reality must prevail during casing design. The logistic implications of mixed grade strings or grades can outweigh any potential economic advantages they might give. Remember that someone has to run the casing into the ground at the end of the day and it is in everyone's interest to try to keep that part of the operation as simple and trouble-free as possible.

Reservoir fluid gradient

The fluid within the reservoir and therefore its gradient has a direct bearing on the burst mode calculations, and it is very important to research the information carefully. A gas gradient is always used in the reservoir, its value being the minimum possible, unless good information from the production testing of offset wells indicates otherwise. Minimum values are unlikely to be less that 0.14 p.s.i./ft (0.32SG). The reservoir gas gradient should also be used in larger diameter hole sections above the reservoir. This makes the worst case assumption that gas enters the higher section by migration or an inadequate seal. (It is only the fluid gradient which is assumed to be the same and not reservoir pressure.)

Casing shoe setting depths

The general criteria for the selection of casing shoe setting depths must be that the preceding hole section can be drilled successfully and safely. The casing shoe must be set in competent formations that will be able to withstand the forces imposed upon it should a pressure kick occur.

The shoe depth selected for the conductor pipe should be such as to provide a shoe strength strong enough to withstand fracturing during drilling the next hole interval which is assumed to have no hydrocarbon intervals. To estimate the anticipated fracture pressure, the following conditions must be considered:

drilling rate with loading effect in annulus
equivalent circulating density
mud weight to be used

Surface casing is treated as for the conductor pipe if no hydrocarbons are expected in the next hole interval or alternatively as for the intermediate and production casing, in the event that hydrocarbons can be expected.

The shoe depth selected for intermediate and production casing should be such as to provide a shoe strength strong enough to withstand fracture during drilling the next hole section, according to the criteria established for the conductor pipe and also strong enough to take a gas kick of defined volume.

The mechanisms for selecting a shoe depth are as follows:

1. Clearly define well objectives.
2. List actual problems encountered in nearby wells.

3. List potential problems encountered in nearby wells.
4. Estimate pore and fracture pressure profiles for well.
5. Overlay pore and fracture pressure profiles with expected well lithology targets and objectives.
6. Make a basic casing design (try four strings).
7. Study production casing, shoe depth requirements to satisfy designed kick tolerance.
8. Select suitable formation and depth to meet this requirement as an absolute minimum.
9. Recalculate kick tolerance for selected shoe depth.
10. Study intermediate casing shoe depth requirements to satisfy designed kick tolerance.
11. Select suitable formation and depth to meet this requirement as an absolute minimum.
12. Recalculate kick tolerance for selected shoe depth.
13. Study surface casing shoe depth requirements to satisfy next hole section requirements.
14. Select suitable formation and depth to meet this requirement as an absolute minimum.
15. Study conductor pipe shoe depth requirements to satisfy next hole section requirements.
16. Select suitable formation and depth to meet this requirement as an absolute minimum.

Kick tolerance (KT) is a determination of the size of kick that can be safely shut-in and circulated out of the well. Kick tolerance should be calculated both:

1. At the programming stage for the formation at the chosen casing shoe depth for all casing strings below the conductor.
2. During drilling once a formation leak-off test has been performed use the formation strength figures to recalculate the kick tolerance. Periodically during drilling, recalculate kick tolerance as mud weight changes or potentially weaker formations are penetrated.

Since in most cases, the weakest point in the hole is immediately below the casing shoe, the kick tolerance can be calculated by determining the pressure at casing shoe when the top of the gas bubble is circulated to the casing shoe. The kick tolerance is calculated from the following equation:

$$\mathrm{KT} = \frac{\mathrm{SD} \times \mathrm{SFS} \times \mathrm{Ann}}{\mathrm{PF} \times (\mathrm{MG} - \mathrm{GG})} \times [(\mathrm{SD} \times \mathrm{SFS}) - \mathrm{FD} + \mathrm{MG}\ (\mathrm{TVD} - \mathrm{SD})]$$

where

KT = volume of influx in barrels that can be tolerated
Ann = annular capacity below shoe, hole/DP (bbl/m)
SD = depth to casing shoe (m)
TVD = depth of section (m)

PF = formation pressure at TD (p.s.i.)
MG = mud gradient (p.s.i./m)
GG = gas gradient (p.s.i./m)
SFS = formation breakdown gradient (p.s.i./m)

Design criteria

The following are the criteria which must be considered when carrying out detailed casing design:

1. Burst
2. Collapse
3. Tension
4. Compressional effects

Burst is pipe failure which occurs when the pressure inside the pipe is greater than the internal yield value of the pipe plus the pressure outside the pipe. Burst mode calculations must be made to ascertain exactly what volumes of gas kicks can be taken and at what pressures.

Both the strength of the formation at the shoe, and the burst strength of the casing must be considered. The ideal situation is for both formation and casing to be able to withstand the pressures resulting from a full gas column to surface and the additional pressures resulting from circulating the gas out. However, for deeper wells of a higher pressure, it must be accepted that both formation and casing may have to be designed on a limited kick basis.

It is recommended that the pressure profile chart method is used for the complete burst scenario.

1. To roughly ascertain kick volumes and casing burst values.
2. As a final graphical representation for presentation. Included in the graph will be lines representing porc pressure, fracture gradient, gas gradient, mud gradient, net burst, casing burst and lengths of allowable kicks.

The graphical presentation will show the situations for burst on the formation and burst on the casing.

The Drilling Engineer should ensure that the formation at the shoe is always weaker than the casing. This criteria is in the interests of the safety of both personnel and the rig. If pressures resulting from a gas kick were to rise to an unanticipated level when shut-in, an underground blowout is preferred to the casing failing.

In the design stage every factor involved in the burst on the formation calculations is an estimate based on offset data. As the well is being drilled, these figures should continually be recalculated as more accurate information evolves. At minimum, this should be performed after every leak-off test is conducted.

However, it is good practice to re-evaluate the kick tolerance at more regular intervals using the mud logger's estimates of pore pressure, especially if the pore pressure profile is increasing. Although the formation at the shoe normally has the lowest fracture gradient in the hole section, it should be borne in mind that weaker formations may exist below the shoe. If the well to be drilled is designed with low limited kick volumes (below 100 Bbls) the operator must ensure that the rig, the rig equipment and the personnel he uses are adequate for fast kick detection and briefed fully on the situation and its implications.

When a formation 'kicks', its flow regime is what is termed 'transient'. It does not immediately conform to the standard steady state flow equations used in production technology and is more analogous to production test flow rather than production well flow. The actual inflow performance will depend on:

1. Drawdown (the difference between mud, hydrostatic pressure and formation pressure)
2. Formation porosity
3. Formation permeability
4. Wellbore damage (usually due to drilled solids and filter cake blocking well bore pores)

Formation porosity and permeability can be considered to be constant for a given formation in a given well. Wellbore damage and drawdown will vary, however, during the flowing process.

Solids which have been blocking wellbore pores can be pushed back into the wellbore by the flowing influx, hence reducing damage to the wellbore. Furthermore, as gas is produced into the wellbore to take the place of mud in the annulus, the drawdown on the well increases. The significance of these two phenomena, in practice, is that most wells will kick slowly initially, and with increased intensity until they are closed in. Consequently, early kick detection is paramount for safe drilling operations.

Kick detection is a function of the personnel involved, mainly the Drilling Supervisor who lays down the ground rules and the Driller as the first line of detection. It is also dependent on the quality of the kick detection equipment installed at surface. Mud loggers can also play a vital part in kick detection by setting up their equipment properly and having an intimate knowledge of it. Their instrumentation will quite often pick up tank gains before the Driller's instrumentation does.

For a low kick tolerance well, it stands to reason that the operator should closely consider the above. Nothing can be done about the formation or the fluid characteristics, but the hole diameter and annulus volume are dependent on the casing design. The operator should insist that the rig contractor supplies experienced drillers, with additional training if considered it necessary. Kick detection equipment should be of a high standard and an upmarket mud logging package be procured. Utilising a small active system will assist in kick detection.

If the Well Designer has any comments to be made on any aspects of a small kick tolerance well, including considerations of the above, he must detail it in the casing design. It should be borne in mind that the research and work involved in the casing design implies that that particular Planner is the most knowledgeable drilling person associated with the pressure and strength regime in the well.

The Well Designer should ensure that the casing is stronger than the formation in the burst mode. However, it is accepted that there is the odd occasion where the formation gradient data simply does not allow this. A safety factor of between 10 and 30 per cent is usually applied to the internal yield values of the casing.

The liner should be considered first, then the production, intermediate and surface strings, taking the net burst into consideration in all cases. The net burst is the difference between the pressures inside and outside the casing. If it is greater than the casing's rated burst pressure, less design factor, the pipe is in danger of failing.

The worst condition for the liner or production casing is complete evacuation of the test string or completion to reservoir fluid with a tubing leak occurring at the surface. A combined pressure of the annulus hydrostatic plus the surface pressure is exerted above the packer. The worst situation assumes that only pore pressure is behind the liner. If applicable, other factors such as fracturing and acidising must be considered.

In the interest of economy, the liner is often tied back to surface to allow a lower grade production casing to be used. Assuming it is uncemented, the tied back can be retrieved prior to abandonment. For the intermediate strings of casing, the worst condition is assumed to be complete evacuation to the formation fluid. When circulating out a gas kick, the pressures exerted on the upper section of the casing will be greater than the pressures under a static condition. Pressure testing of casing should also be considered.

Collapse will occur when the external force on the pipe is greater than the combination of the internal force plus the collapse rating. It occurs as a result of either or a combination of:

1. Reduction in hydrostatic head exerted by the fluid inside the pipe.
2. Increase in hydrostatic head exerted by the fluid outside the pipe.
3. Mechanical forces created by plastic formations, squeezing salts etc.

The above three factors can result from the following situations:

1. Inadequate fill up of casing when running
2. Lost circulation
3. Cementing
4. Casing wear (see page 120)
5. Air or foam drilling. The casing has to be designed for complete evacuation plus an allowance for wear due to loss of lubricity
6. Halite sections
7. High drawdowns for testing purposes. It is generally accepted that a rank exploration well will not be subjected to high drawdowns but this should be considered for appraisals.
8. Acidising of fracturing a horizon could result in an increase in external loading to a depth above the packer if a path of communication exists.
9. Similarly squeeze cementing could increase external loadings above or below packers.
10. Corrosion will eventually decrease the collapse strength of the pipe.

For most wells, only the first three situations are usually considered. For the first two, a good indication of resistance to collapse is the percentage of the casing which could collapse, either due to lost circulation or not filling the pipe when running. It should be mentioned that the latter is strictly an operational consideration and not a design one. As such, it should not be a design criteria. The result of the two situations is simply stated as a percentage and no design factor needs to be incorporated. The operator may wish to stipulate a value of say 40 or 50 per cent, being the minimum proportion of casing to be full at all times.

A second method of considering the lost circulation situation is to assume the following worst case. When drilling the next hole section with the maximum allowable mudweight, total losses are experienced. This would result in the mud level inside the casing dropping to an equilibrium level where the mud hydrostatic equals the pore pressure. The operator may wish to stipulate a design factor of between 1.0 to 1.15.

In the third situation, the various cementing possibilities should be considered, to ensure that the heavy cement outside coupled with spacers or light weight mud inside does not induce collapse. Again, a design factor of between 1.0 and 1.15 is usually imposed. Large diameter casings generally have a low collapse resistance. This disadvantage is made worse by the fact that they are often cemented by inner string techniques. Annulus pressures and the possibility of packing off have to be looked at.

For the collapse design of casing strings which are set deeper (say below 12 000 ft), in high mud weights (say above 14 ppg) or have high rates of bending (say above 10 degrees/100 ft), it is recommended that biaxial forces be taken into account (see page 120).

Tensile failure will occur if the pull exerted on the pipe is too great for the tensile strength of the pipe or coupling. With the 'off the shelf' string of casing, the coupling is stronger than the pipe (although a check should always be made). That criteria should be specified when ordering from the suppliers. For situations where the coupling has to be special clearance, for example of a smaller diameter than the normal, the coupling will be weaker. The number of parameters which can affect tension means that the estimates for the tensile forces are more uncertain than the estimates for the burst and collapse. The safety factors imposed are therefore much larger, being in the range 1.30 to 1.80. Tensile loads on the casing should be calculated at the following stages:

1. When running the pipe
2. When cementing
3. When pressure testing (drilling phase)

Four situations should be considered and a safety factor evaluated for each one.

Situation 1. **Common force** plus a **shock loading** when running
Situation 2. Common force plus an **overpull** when running
Situation 3. Common force plus a **weight of cement** force when cementing
Situation 4. Common force plus a **pressure test** in the drilling phase

The highlighted terms are explained in more detail below.

Common force is a combination of the weight of the casing string less the buoyancy force in the minimum mudweight envisaged plus a bending force.

(i) Weight of the casing (Wt.)

$$Wt. = W \times L$$

where
W = casing weight in lb/ft
L = casing length in ft

(ii) Buoyancy force (BF) is an upward force acting on the bottom of the casing string. True vertical depths must be used in a directional well. Any composite strings with different ID's must be considered separately.

$$BF = MW \times CSA \times L$$

where
MW = mudweight in p.s.i./ft
CSA = cross sectional area in sq ins
L = casing length in ft

(iii) Bending force (BeF) is a force acting in tension on the outside of the pipe and in compression on the inside. It will be caused by any deviations in the well, resulting from side-tracks, build-ups and drop-offs or from sagging of casing caused by lack of centralisation or washouts. A radius of curvature of 1 or 2 degress/100 ft is normally used for vertical holes and 5 degrees/100 ft for build up or drop off sections. (It should be noted that a survey may not indicate the real rate of curvature in a dogleg if the stations are not close enough together.) Bending calculations must be re-done if a well has to be side-tracked round a fish.

$$BeF = 63 \times RC \times D \times W$$

where
63 = a constant incorporating CSA, W and Youngs Modulus for steel
RC = radius of curvature in deg/100 ft
D = outside diameter of pipe in ins
W = casing weight in lb/ft

Shock loading, when running, is exerted on the pipe because of:

1. Sudden deceleration forces, for example if the spider accidentally closes, or the slips are 'kicked-in' on moving pipe, or the pipe hits a bridge.
2. Sudden acceleration forces, such as picking the pipe out of the slips, or if the casing momentarily hangs up on a ledge then slips off it.

Any of the above will cause a stress wave to be created, which travels through the casing at the speed of sound.

$$\text{Shock loading} = 150 \times V \times CSA$$

where
150 = the assumed speed of sound in steel in lb sec/in
V = peak velocity when running in in/sec
CSA = cross sectional area in sq ins

Overpull contingency of 100 000 lbs is normally incorporated. This is not exactly a design factor but a function of the hole conditions.

Cement force (CF), a worst case situation is assumed as follows: the mudweight in the annulus is the lowest envisaged for the section; the inside of the casing is full of cement slurry, with mud above; the shoe instantaneously plugs off just as the cement reaches it and the pressure rises to a value of say 1000 p.s.i. before the pumps are shut down. It is appreciated that the cement will be 'running away' at this point with no positive displacement pressure being exerted.

$$CF = [(CW - MW) \times L + 1000] + A$$

where
CW = cement weight in p.s.i./ft
MW = mud weight in p.s.i./ft
L = length over which CW & MW act in ft
A = internal area of the casing in sq ins

Pressure testing will be performed on the casing as the plugs are bumped and later on in the well depending on operational conditions. The actual test pressure will depend on:

1. The rated burst strength of the casing
2. The well head pressure rating
3. The BOP stack pressure rating
4. The maximum anticipated surface pressure

Compressional effects occur in casing due to temperature effects in landed casing and because of the weight of other inner casing strings which must be supported by the outer strings. In most cases the outermost casing will carry some or all of the following compressive loads:

the weight of the intermediate casing strings
the weight of the well head
any completion tubing tension

So far as compression loads are concerned, wells fall into one of three categories:

1. Land wells and subsea wellheads
2. Platform wells with surface wellheads
3. Mudline suspension wells

In land wells, provided the outer casing is cemented all the way to surface it will be able to support all the expected compressional loads. If, however, it is not cemented to surface then there is a danger of buckling due to the compressive loads.

With platform wells with surface wellheads, there is a freestanding part of the casing equivalent to the water depth plus air gap plug height to the wellhead deck. Buckling can occur on this freestanding section. To prevent buckling, the outermost casing must be well centralised within the diver conductor and designed to be strong enough to withstand the likely buckling forces.

With mudline suspension wells, used mostly on jack-up wells, the weight of the casing is hung off at the seabed. The tieback strings which link the seabed wellhead with the surface equipment on the jack are however, subject to buckling. Except in areas of high seas or current a 20 inch surface casing string of 133 lb/ft will be sufficient to support surface equipment in water depths up to 90 m.

During drilling operations, temperature effects are so slight that they can be ignored. During the production phase, however, the compressive loads on the production string must be considered, especially if the well is to be used as an injector. A design factor of 1.0 should be applied in these cases.

Special considerations

Liners have to be considered in a wider sense, in addition to the criteria outlined on page 113, because of its function, method of setting and the formations in which it is normally set.

1. Reciprocation

The cement bond is of prime importance to provide isolation of different horizons within the hydrocarbon bearing formations. To give the best possible chance, the operator may want to reciprocate or rotate the liner. With reciprocation, the rig personnel must be aware of the maximum overpull permissible. Tension calculations should be reconsidered for this situation. On a heaving floating rig, reciprocation is unavoidable until the casing is landed.

2. Position of packer

The likely position of the production test packer has a bearing on the burst design. The packer can be set in three positions, dependent on the testing requirements and the depth from the last casing shoe to TD:

(a) inside the production casing
(b) inside the liner overlap
(c) in the liner

With situation (a), the complete liner would be subjected to the formation pressure. With situation (b) and (c), the liner above the packer is subjected to the pressures resulting from a tubing leak at surface as outlined on page 113.

3. Tied-back liner

For a high pressure well, the intermediate string may simply be unable to withstand the tubing leak at surface pressures, acting in the burst mode. The solution to this problem is to run a full string of pipe to surface. Within the full string choice are two options. Firstly, to run a full string of casing, and secondly, to run a liner and tie it back.

The following facts should be considered:

1. Running a full string is easier, faster and hardware can be cheaper.
2. If, when running a full string, the pipe gets stuck, it is difficult to recover from the situation, probably resulting in a cut and patch operation which will downrate the casing design immediately. Running a liner would at least allow the well to be evaluated by setting the hanger high and running a contingency string of small liner through it.
3. Circulating pressures when cementing will be lower with a liner option.
4. With the liner option the annulus can be mechanically sealed off if the cement job is not adequate.
5. The forces resulting from pressure and temperature on a string anchored at top and bottom can be negated by an uncemented tieback, where a mandrel slides in and out of the Polished Bore Receptacle. (PBR)

If the tieback option is chosen, in addition to the liner being cemented, the tieback can be cemented. However, with an anchored string, the buckling, ballooning and positioning forces resulting from pressure and temperature are much more difficult to calculate than the resultant length changes in an uncemented tieback.

Biaxial forces may be neglected in the design of casing strings unless:

1. Casing is being set deeper than 4000 m
2. Mud weight is greater than 1.7 SG
3. Bending rate is in excess of 10°/30 m

Casing wear decreases the performance properties of casing. The burst and collapse resistance of worn casing is in direct proportion to the remaining wall thickness. In practice, casing wear is proportional to some or all of the following:

1. Rotation of BHA and drillpipe
2. Tripping
3. Inclination of hole
4. Rig positioning over well
5. Severities of dog-legs

6. Solids content in mud
7. Type of mud (lubricity)

If there are worries about casing wear, it stands to reason that operational practices should be used that minimise wear. These include:

1. Use of downhole motors and turbines
2. Use of rubber DP casing protectors*
3. Use DP without hard banding or with flush smooth ground hard banding
4. Keep dog-legs to a minimum
5. Keep sand content low
6. Use of oil-based mud with good lubricity

Detecting casing wear can be done by two methods:

1. Use of magnets in mud flow return to measure the amount of steel cuttings returned in the mud flow.
2. Running a capiler survey after setting the casing to provide a base log. A wear log can be run at any time throughout the life of the next section.

As the reduction in burst and collapse properties is directly proportional to wall thickness, the revised theoretical value may be calculated. There is no reliable method of predicting casing wear and defining the reduction in that casing's properties. The normal procedure to cater for possible wear when designing, is to move up a casing grade or have a greater wall thickness. In vertical wells, casing wear is usually in the first few joints below the wellhead, or in intervals with a high dogleg severity. Consideration should be given to increasing the grade, or wall thickness of the first few joints below the wellhead. In deviated wells, wear will be over the build-up and drop-off sections. Again the casing over these depths can be of a higher grade or greater wall thickness.

Flow salt zones are occasionally encountered but are common in the southern North Sea. They can cause unexpected collapse of casing since their effective overburden stress gradient is greater than normal. A figure of 1.25 p.s.i./ft should be assumed for collapse calculations.

Corrosive fluids should be contained within the tubing. They should not be produced through or injected into the casing/tubing annulus. However, it is accepted that tubing leaks and pressured annuli are a fact of life and as such production casing

---

*There are two schools of thought on the merits of using rubber DP casing protectors. The author is in favour of not using them at all and relying on flush smooth ground hard banding to limit casing and drillpipe wear. The thought behind this is that the rubber protector can become a soft matrix into which can be embedded hard formation particles. This creates a grindstone type of composition which can damage the casing more than the drillpipe tool joint would. In true drilling fashion, however, not everyone agrees with the author on this matter so in most cases you will have to follow your own instincts or your company's drilling policy, if it has one.

strings are considered to be subject to corrosive environments when designing casing for a well where hydrogen sulphide or carbon dioxide laden reservoir fluids can be expected. During the drilling phase, if there is any likelihood of a sour corrosive influx occurring, thought should be given to setting a sour-service casing string before drilling into the reservoir. The BOP-stack and wellhead must also be suitable for sour-service.

For Exploration/appraisal wells, routine measures taken during drilling include:

1. Using casing and wellhead equipment with a metallurgy for sour-service
2. The use of high alkalinity mud to neutralise hydrogen sulphide gas
3. The use of inhibitors and/or scavengers

These measures will provide a degree of short-term protection necessary to control corrosion of the casing in the hole.

For development wells casing corrosion considerations are restricted to the production casing.

Internal corrosion can be avoided if the well is designed to contain any corrosive fluids (produced or injected) within the tubing string by using premium connections. Any part of the production casing that is likely to be exposed to corrosive environment during routine completion/workover operations, or in the event of a tubing or wellhead leak, should be designed to withstand such an environment.

Where the likelihood of external corrosion due to electrochemical activity is high, and the consequences of such corrosion are serious, the production casing should be cathodically protected (either cathodically or by selecting a casing grade suitable for the expected corrosion environment).

Casing to be used for sour service must be specified according to the API 5 AC for restricted yield strength casings. The casing should also meet the following criteria:

1. The steel used in the manufacture of the casing should have been quenched and tempered. This treatment is superior to tubulars heated/treated by other methods such as normalising and tempering.
2. All sour service casing should be inspected using non-destructive testing or impact tests only, as per API Specification 5AX.
3. Downgraded grade N80, P105 or P110 tubulars must not be included in orders for K55 or J55.
4. The coupling must receive the same heat treatment as the pipe body.
5. Cold die stamps are prohibited; all markings must be painted-stencilled or hot die stamped.
6. Extreme care must be taken when handling sour service tubulars. Every notch in the wall is a potential stress riser that could cause the pipe to fail.

The existence of any of the following conditions alone, or in any permutation, may lead to the initiation and perpetuation of corrosion:

1. Hydrogen sulphide, sour gas, or sour oil and gas multiphase
2. Sour brine

3. Carbon dioxide gas
4. Water
5. Temperature
6. Velocity of fluids within the environment (erosion)
7. Dissimilar metals, galvanic corrosion

The various types of corrosion are briefly summarised below. As far as casing corrosion is concerned, the three types of corrosion that must be controlled are:

1. Sulphide stress corrosion cracking
In a sour environment, brittle failure usually occurs by cracking. This takes place due to the combined action of tensile stress and corrosion, in the presence of water and hydrogen sulphide.

2. Chloride stress corrosion cracking
This is corrosion in a high chloride environment and usually takes the form of highly localised anodic dissolution of the metal under tensile stress. The strength of the material has little or no effect on the susceptibility of materials to this form of attack. It is important to know that chloride can and does promote sulphide stress cracking in corrosion resistant alloys. Here the chloride reduces the effectiveness of the alloy's passive film and thus encourages corrosion. This increases the hydrogen evolution rates and more hydrogen will enter the alloy to cause embrittlement.

3. External electrochemical corrosion
Electrochemical corrosion of casing strings occurs on casing strings set in formation with low or no electrical resistance, and in salt or salt water bearing formations. This type of corrosion requires electrical current to flow from the casing to the formation and may be very difficult to detect.

Buckling is a stability failure and can occur at stress levels well below the yield stress of the material. Buckling does not occur where the casing is supported by cement. Factors responsible for buckling and the degree of buckling are:

1. Length of casing, supported by cement
2. Hole size and degree of washout
3. Tensile loads on the casing string
4. Increased mud weight or pressure from other sources that increase internal casing pressure
5. Temperature increases downhole

All these factors are interrelated, but the first three are generally considered to control the degree of buckling, while temperature and pressure changes are primarily mechanisms that cause the initial buckling.

A buckling potential may exist in the uncemented portion of a string of casing, if, after the cement has set:

1. The internal mud density is increased
2. The internal surface pressure is increased
3. The annular fluid removed or its density reduced
4. The casing is landed with less than full hanging weight
5. The temperature of the casing increases

Buckling of long, uncemented portions of the casing string, in 'vertical' wells, can be prevented by:

1. Cementing the casing up to the neutral point
2. Pretensioning the casing on landing
3. Limiting the increase in mud density used after drilling out the casing
4. Rigidly centralising the casing below the neutral point

Provided that all casing strings are landed with full hanging weight, buckling calculation can be limited to the small percentage of deep vertical wells in which the mud density is to be raised during the drilling of the next open-hole section. Thus, for the majority of wells, buckling is not a major design consideration.

Casing centralisation

The centralisation of casing and liners and also the correct positioning of the centralisers is an essential part of any casing or liner job. Without adequate centralisation it is possible that the casing can make contact with the wall of the hole preventing effective annular sealing by the cementation. Centralisers should preferably be pre-installed on the rack to save rig time.

Two basic types of centraliser are used:

1. Spring type centralisers
   (a) Large annuli: Centralisers are installed over a stop collar and should be of the Weatherford St-II, ST-III or St-IV type or similar.
   (b) Close tolerance annuli: Centralisers are installed in between stop collars such that they can fully collapse. In the case of a collared coupling, the upper stop collar is replaced by the coupling. Centralisers should be of the Weatherford ST-1 type or similar.
2. Rigid centralisers are used inside casing only. They are installed over stop collars, or in between stop collars or collared coupling and stop collar. The method used depends entirely on the centraliser design.

Centralisers should preferably be fixed over a stop collar. However, it is necessary to fit stop collars above and below the centralisers when using spring type centralisers of close tolerances, or when using rigid centralisers which do not have space to straddle a stop collar. If the centraliser can be pre-installed on the rack this should be done to save rig time. If pre-installation is not possible and when using collared type couplings, centralisers can be installed over the coupling.

The suggested centralisation of various casing string sizes are outlined below:

20″ and 18⅝″ casing The centralisers should be placed as follows:
Two spring centralisers on the shoe joint, one 5 ft above and one over or below the collar.
One spring centraliser on the 2nd and 4th joint.

13 ⅜″ casing The centralisers should be placed as follows:
One spring centraliser 5–10 ft above the shoe.
One spring centraliser on the 2nd, 3rd 4th and 5th joint.
One spring centraliser every five joints for open hole angles less than 20 degrees.
One spring centraliser every two joints for open hole angles more than 20 degrees and one centraliser each joint in built up section for open hole angles more than 20 degrees.
One rigid centraliser below the hanger.
One spring centraliser inside the previous casing shoe.

9 ⅝″ and 7″ casing The centraliser should be placed as follows:
Two spring centralisers on each of the first two joints **over** a stop collar for 9 ⅝ inch casing.
Two on each of the first two joints **between** stop collars for 7 inch casing.
One spring centraliser on each joint over build up section.
One spring centraliser every two joints for open hole angle less than 20 degrees.
One spring centraliser every joint for open hole angle greater than 20 degrees.
Two spring centralisers on each joint of 7 inch casing for open hole angles greater than 45 degrees.
One spring centraliser inside the previous casing shoe.
One rigid centraliser one joint below the hanger.

7″ and 4½″ liner The centralisers should be placed as follows (excluding underreamed hole):
Two spring centralisers on each of the first two joints between stop collars.
Two spring centralisers per joint in open hole.
One rigid centraliser per joint over the liner lap.

Running procedures

If the Drilling Operations Manual does not comprehensively cover running procedures, the Drilling Programme should include information such as that given below.

When running casing (or tubing) whether premium or API, be sure to follow the basic recommendations in API RP 5 C1:

1. Check that the tubulars run are clean and have clean, freshly doped connections.
2. Avoid damage by leaving the protectors in place during handling.
3. Use recommended torque during make-up.

4. Always use API modified compound.
5. If no API modified compound is available be sure the proper friction index is used.
6. Be sure the torque gauge is adapted to the power tong torque arm.
7. Check the torque gauge accuracy.
8. Do not weld on heat treated tubulars (i.e. N-80 grade and above).
9. Always use NRV type float equipment not differential fill up type.

The following is a list of preparations for casing:

1. When casing is delivered and racked, remove the protectors and thoroughly clean and check the casing threads. Dope the female threads with the required lubricant, and replace the pin end connectors. (Do not grease pipe in the rotary table). Ensure that the protectors are backed off sufficiently to obtain the proper measuring point.
2. Accurately measure, drift and number the casing on the rack as it arrives on site. The casing must be measured independently by two sources under the direction of the DS. After a layer is completed, the DS should check both pipe tallies and ensure that there are no differences in lengths. In case of such differences the pipe(s) should be re-measured by the DS. Defective joints are to be clearly marked as such and have to be removed to a separate area.
3. When measurement is completed the DS should check the total amount of joints with the delivery note. As an additional check, he should calculate the average joint length.
4. The DS should then make up the casing running programme and fax a copy to base. Joints to be excluded from the string should be clearly marked and the running list should indicate clearly the joints where centralisers should be attached. Centralisers should preferably be placed on the pipe body on the rack and not over collars because this slows the running of the pipe. Check with the caliper log. Copies of the running list should be handed out to all personnel involved with running the casing, such as the contractors TP, Driller, Crane Operator, Roustabout Foreman and Sample Taker.
5. Check replacement of shoe and collar or special casing equipment. **Welding on casing in not permitted**. The collars of the joints making up the shoe track should be removed and made up again with the proper torque and a thread-locking compound should be used. Flat collar, shoe and DV collars should be made up similarly.
6. The casing tally can be checked by multiplying the number of joints run in the hole by the average length as calculated in 3 above. Any significant difference in the resultant casing setting depths calculated by the two methods indicates an error which should be checked.
7. As soon as the casing is about to be landed, the DS should count the joints of pipe remaining at that stage and check this against the total amount of pipe received. This is to ensure that the correct length of casing is run. This also applies when a DV collar (or any other special joint or piece of equipment) is made up. All such accessories should be made up on casing pin ends to ease fishing should the casing drop through the slips into the hole.

8. Ensure that sufficient water and displacing fluid are available and that adequate supply lines are provided for the cementers.
9. Determine the amount of cement required; from the caliper log or experience, and the Drilling Programme.
10. Order cement and additives well in advance and notify cementers in time to have the cement equipment rigged up prior to landing the casing.
11. Check that the mud is within specification prior to pulling out of the hole for casing. Also ensure that the displacement fluid is pumped and received into separate tanks, such that a positive method for measuring both displacement volume and lost volume is available.
12. Mud pumps should be fitted with the proper size liners and be in good mechanical condition. Have a mud fill up line to the casing rigged up with a quick opening valve.
13. The hole depth must be checked by strapping out of the hole at least once prior to reaching a critical depth, such as the casing point, logging point etc., and if these measurements do not agree with the driller's depth, the pipe should be re-measured. Also check the depth on the geolograph.
14. If a liner is to be run, hook up lines for reversing and make sure that the amounts of drillpipe on the rack and in the derrick are known.

The following is a checklist for running casing:

1. Visually check the inside of each joint of casing on the rack to see that all joints are clear of foreign matter. Make up the float collar on the bottom of the next joint before running it, so that if the pipe falls through the slips, a spear can enter into the top of the joint. (This is easier than having to deal with a float collar at the top of the joint.) Both the shoe joint and the collar joint should be fitted with a blanked-off casing thread protector as soon as the shoe and collar have been installed. This is to prevent foreign objects from entering either joint during storage and handling.
2. On liner jobs and 20 inch casing cementations drift all drillpipe used with the proper size 'rabbit' and see that the rubber wiper plug is the correct size for the drillpipe.
3. Always place a short joint in the casing string near the pay zone to aid later in checking depths with a casing collar locator (CCL).
4. Check conventional casing float equipment and surface mud lines after the shoetrack is run in.
5. Fill up the casing every joint and completely every five joints.
6. Ensure that the casing is properly made up.
7. Run the casing smoothly. Avoid high acceleration and deceleration which cause high surge/swab pressures. Regulate the casing, lowering speed to 30 sec/jt (1.33 ft/sec). Caution the driller about possible lost circulation if the casing is run too fast. Check returns every joint and where possible use the possum belly tank to fill up the casing.
8. When the side door elevator is in use, avoid impact loading which can open this type of elevator. Particular care must be taken when centralisers are entering the

rotary or BOP/wellhead. When the hook load approaches casing pipe and collar strength and when leaving casing shoes, use slip type elevators (not for 20 inch casing).

9. Check the cementers hook up, have the cement unit and manifold ready for use. Pressure test and flush the lines prior to cementing.

10. Visually check to be sure that the correct bottom plug is properly placed in the cement head prior to circulating the hole after landing the casing. On a two-stage cementation, load the shut-off plug of the first stage if possible. After dropping the DV opening bomb or plug, load the closing plug in the cement head.

11. See that a correct size shear pin has been installed in the relief valve of the rig pump or check the fitting of the relief valve.

12. Using the rig pump break circulation slowly. Record the free hanging weight of the string.

13. Circulate at least the casing contents + 20 per cent. Record the circulating pressures and rates up to the maximum rate allowable from pressure considerations, i.e. 85 per cent of formation breakdown, casing test pressure, losses etc. Pump and displace cement at moderate rates compared to the maximum indicated above. In the case of 9 5/8 inch casing aim to achieve a 250 ft/min annular velocity. In the case of 7 inch casing, 350 ft/min.

In the case of gas cutting, circulate (increasing mud weight where necessary) until the gas cutting has ceased and there are no further gas readings. During circulation, stop the pumps to check if gas movement is still present under static conditions.

14. Precede the bottom plug with a preflush and place the top plug in the cementing head after releasing the bottom plug. The amount of preflush depends on hole conditions, hole size and mud type and will be specified in the well programme.

15. Mix cement to the required slurry weight and check the weight regularly.

16. Collect numbered samples of the slurry and of at least 10 lbs of the dry mix. Keep them until the results of the cement job have been determined. Also take a water sample and note the temperature when taken. Install a pressure recorder with the lowest acceptable pressure range on the cement pump.

17. Leave the mixing tub full of the required cement slurry at the conclusion of mixing to avoid the possibility of pumping diluted cement or even water into the casing before the top plug is released. Pump this away and flush out the equipment with water.

18. Release the top plug. For pipe sizes of 5 inches or less, do not pump excess cement from the lines on top of the plug. Chase with string suspended from hook (unless BRX used).

19. Do not reciprocate casing strings unless requested to do so by Base.

20. Displace the cement with mud using the mud pump. Where the reservoir pressure is hydrostatic or below displace the cement with water). Determine the displacement volume, pump strokes and pumping time at which the displacement rate should be reduced prior to bumping the plugs. Displace cement at the maximum rate allowable from (m) above unless advised otherwise. It may be considered in some special cases that plug flow is advantageous. When the number of pump strokes required to reach the bumping point (assuming 100 per cent efficiency) has been reached, cut out one pump, or if one pump only, reduced the rate and calculate the additional time for the plug to reach the float collar.

Continue pumping until the plug is bumped unless there are grounds to suspect something is wrong. Remember that mud is compressible due to entrapped air and that the bump can sometimes take place later than expected.

21. Displace from the cementing unit only in the following cases:
(a) When no rig pumps are available. Control volume by measuring from the mud tanks as well.
(b) When carrying out plug backs.
(c) Liner cementation.
(d) Poorboy job on large diameter casing, where no plugs are used.
(e) Duplex cementing with DP stingers.

22. If the rig pump is used for displacement, have the cementer's pump unit ready to take over to bump the plugs in case the pressure becomes excessive for the rig pump. Record all mixing, displacing, bumping, opening/closing of DV collars etc, on charts.

23. Release the pressure, measure returns and check for backflow. If there is backflow, pump back **the amount of backflow only** and repressure the casing. If there is still backflow, wait until cement is hard before repeating the test.

24. Pressure test casing/liner immediately after the bump.

25. If float equipment fails and/or pressure is held on the casing, a pressure gauge should be installed on the cement head so that the required pressure can be maintained and excessive pressure be bled off periodically. In this case, the pressure left on must not exceed the theoretical or observed differential pressure between mud and cement.

Presentation in drilling programme

Having gone through the calculations for the casing requirements for the planned well, the results only should appear in the Drilling Programme. They can be presented in the following manner:

Casing (inches)	Weight lbs/ft	Grade	Setting depth (feet)	Test pressure (p.s.i.)
20	133	K55	1000	500
13⅜	68	K55	3000	1500
9⅝	47	N80	8000	3000

SECTION 4: WELLHEAD SELECTION

Having completed the casing design, we have all the information required to allow us to select a wellhead. The wellhead must be of the correct pressure rating, designed for the desired service (H_2S or whatever) and be capable of accommodating all designed and contingent casing sizes.

In practice this comes down to a choice between a few major manufacturers' products. The final selection will probably come down to the ease of operation, cost and the operator's personal preferences.

Having selected a wellhead, its specification should be included in the Drilling Programme along with a sectional view of its component stack up.

SECTION 5: BOP REQUIREMENTS

The BOP requirements for a given well will depend on company policy and anticipated bottom hole pressures. Surface holes have either no BOP requirement, or will need to use a divertor. This will be specified in this section. The information should be presented in a format such as:

Hole size (inches)	BOP requirement	Rating (p.s.i.)
26	nil	
17½	nil	
12¼	2×rams	10 000
	1×shear	10 000
	1×annular	5 000

SECTION 6: CEMENTING PROGRAMME

Cement is used for zonal isolation in the well. The effectiveness of this zonal isolation depends on three main factors all of which must be considered at the planning stage.

1. Slurry design
2. Casing accessories selection
3. Displacement methods

Slurry design

Oilwell cement slurry consists of cement, water and additives. The first calculation that we must make in cement programming is to calculate the volumes required. This is proportional to the height to which the cement will be displaced to in the annulus. The main types of cement with their depth rating is shown below:

Type	Depth rating
Class A	0 – 1800 m
Class B	0 – 1800 m
Class C	0 – 1800 m

Type	Depth rating
Class D	1800–3600 m
Class E	1800–4300 m
Class F	3000–4900 m
Class G	0–2400 m
Class H	0–2400 m

Secondly, we must know or estimate the bottom hole temperature so that we can predict slurry behaviour downhole. Thirdly, we must look at the mud which is in use whilst drilling and consider what effects, if any, it could have on the cementing process. Fourthly, we look at the formations in which the casing is being cemented to see if these present any particular problems. Finally, we look at the hole geometry to see what annular space consideration might affect the quality of the cement job.

Listed below are definitions of slurry properties.

Yield The yield of the cement, in cubic feet per sack, is the volume of space the slurry will occupy after having been mixed with the water and additives as per design specifications. A high yield slurry is one that will contain relatively large amounts of water and therefore be of relatively low density.

Density This is an important factor since the slurry must be dense enough to control formation pressures but not cause formation break-down. Lightweight slurries can be produced by adding extenders such as bentonite to the slurry mix water or by aerating the slurry. Weighting materials such as ilmeninte, hematite and barite can be added to increase slurry density.

Thickening time This determines how long the slurry remains pumpable and depends mainly upon the cement composition and downhole temperature. It can be manipulated by using retarders.

Mix water The amount of mix water used will depend on the programmed slurry density. The water type used is usually that which is most readily available, i.e. seawater when offshore and freshwater when onshore. If casing is bcing cemented across large or salt or halite sections, then the mix water should be salt saturated to prevent ionic exchange between the slurry and formation.

Compressive strength Cement must have compressive strength to support the casing and isolate zones. The speed at which cement slurry develops compressive strength depends to a great deal on downhole temperatures. If temperatures are too low to allow a particular slurry to reach sufficient compressive strength in a reasonable amount of time, chemicals can be added to the slurry. These accelerate hardening and promote strength development. Calcium chloride is the most common accelerator.

Fluid loss This is the amount of water lost from the slurry to the formation for a given pressure differential. If this is not controlled, then fluid can be lost from the slurry causing premature or 'flash' setting. To combat this problem, fluid loss chemicals are used in the slurry for jobs which will entail high pressure differentials between slurry hydrostatic pressure and formation hydrostatic pressure. Clearly the porosity of the formation will have a big influence on the need for fluid loss chemicals too.

Flow characteristics Good flow characteristics contribute to good formation bonding and less pumping difficulties. Additives are available to reduce cement friction and to encourage slurry turbulence at low pump rates.

Casing accessories selection

Having designed the casing requirements for the well, we need now to look at the accessories that we will use to ensure good cementations. The main accessories used are listed below:

A casing shoe is screwed or welded onto the lower end of the lowest joint of casing. They come in four types:

1. The guide shoe is the simplest. It merely serves as an aid to guiding the casing in the open hole and has a hole in the middle which allows the mud to pass freely into the string. The guide shoe is not widely used.
2. The float shoe is similar, externally, to the guide shoe but contains a non-return valve that allows fluid to flow out of the casing but not into it. The main reason for using this type of shoe is that it gives better well control during running casing. It is difficult to stab in a circulating valve quickly on casing and the float shoe precludes this need if the well were to flow. It means that casing must be filled from surface during casing runs.
3. A differential fill-up shoe uses a valve which allows mud to flow into the casing to keep it full. A steel ball can be dropped and pumped into the shoe to shear out a seat and convert the shoe into a regular float type.
4. The automatic fill shoe is also designed to avoid having to fill up the casing all the time from the surface. It incorporates wedges to hold the check valve open so that the mud can flow both into and out of the casing via the shoe. When required, the automatic fill shoe can be tripped to act like a regular float shoe by pumping down the string at a preset rate to pump the wedges out. Company policy will dictate the type of shoe that must be used on a given well configuration. Float shoes are the most commonly used type.

Collars are usually run one or two joints of casing above the shoe. They serve as a stop for the cement wiper plug so that all the slurry is not displaced into the annulus. Float type collars provide backup for the float shoe. Differential type float collars are available and should only be run in conjunction with a differential float shoe. A baffle collar is merely a ring, usually set in a casing coupling, on which a wiper plug is designed to seat and seal to allow pressure testing of casing.

Note: The pressure rating of shoes and collars must be at least the proposed test pressure of the casing.

Multistage collars are used to locate two distinct speparate columns of cement in the annulus when one continuous column would produce too much hydrostatic pressure on the formation, or specialised zonal isolation, or communication is required.

After performing the first stage cement job an opening 'bomb' is allowed to free-fall and seat in the opening sleeve of the multistage collar. The casing is then pressured up to shift this sleeve down and open the circulating ports. When the second stage slurry has been mixed the wiper plug is pumped down and seated in the closing sleeve. A pressure increase then moves this sleeve down to cover the ports and renew casing integrity.

Centralisers are placed on the outside of the casing to keep it concentric with the hole. Centralisers come in two types, spring and rigid. Rigid centralisers are used just below wellheads and just inside casing shoes. All other applications use spring centralisers. Centralisers are located over or between stop collars which are secured to the casing to ensure that they do not move. (See also Chapter 7, page 124.)

Metal petal baskets are used when there is a risk of the cement slurry being lost to a weak formation and use of a multistage collar is precluded. They are attached to the outside of the casing and are designed to help support the column of cement and prevent it from moving downwards before it sets.

Plugs have a hollow drillable alloy interior and rubber outer fins to wipe clean the casing wall. The bottom plug has a diaphragm that is ruptured by pressure after it seats on the collar. The top plug has a solid alloy insert and is used to pressure test the casing against cementing. Plugs come in sizes which cover various weights of a given casing size. When ordering them, the casing weight and size must be stipulated.

Displacement methods

The most commonly used displacement methods are:

1. Single
2. Multistage
3. Stinger
4. Liner cementing
5. Subsea launch

Single displacement is the most widely used form of cementing and is usually carried out as listed below:

1. After landing the casing, circulate 150 per cent contents to ensure that there is no crud to block the floats, the hole is clean of cuttings and the mud out is uniform throughout the hole.
2. Hook up and pressure test the cement lines to final selected test pressure.
3. Drop the bottom wiper plug.
4. Pump a preflush fluid or scavenger to prevent mud/cement contamination and help remove mud cake.
5. Mix and pump the designed slurry.
6. Drop the top wiper plug.
7. Displace the cement.
8. Bump the cement plugs.
9. Increase the bump pressure to the casing test pressure.
10. If float equipment has been used, bleed off the test pressure and check for back-flow.

Multistage displacement proceeds as above for stages 1. to 5., however, after the initial slurry has been pumped, a shut-off wiper plug is dropped which lands on the shut-off baffle plate. An opening bomb is then dropped which opens the stage collar. Then 150 per cent pipe contents are displaced through the stage collar ports, followed by the calculated cement slurry. Behind the slurry is pumped the top closing plug which closes the stage collar ports allowing the casing to be pressure tested.

Stinger technique is used only on large diameter casing. The casing is landed in the normal manner. A stab in sub which matches the float shoe in the casing is made up on drill pipe and a bow spring centraliser placed just above the sub to facilitate easy stabbing in. Sometimes, heavy weight drill pipe has to be used to overcome the pump out forces which are created by the differential pressure between the fluid column in the drillpipe compared with that in the annulus between the casing and open hole.

Liner cementing requires great care to do it well. The liner is run in on drillpipe and suspended using a liner hanger, then cement is pumped down the drillpipe and displaced with a wiper dart. This wiper dart latches into a wiper plug shear-pinned below the liner hanger and together they displace the cement from inside the liner. The main problem in liner cementations stems from the small annular volumes between the liner and open hole. These manifest themselves as follows:

1. Small cement volumes required mean that accurate displacement is vital or the cement may not end up where you want it.
2. Small cement volumes also mean that cement contamination can occur much more readily than with large jobs.

Once the plug has been bumped and the float equipment operation confirmed, the running tool should be released and mud circulated down the drillpipe to remove any excess cement from around the top of the liner.

Subsea launch techniques are used on floating rigs since these heave up and down with wave and swell and they tend to be equipped with subsea wellheads. In subsea wellheads, the wellhead is latched into the permanent guide base which sits on or just above the seabed. The weight of the casing is therefore taken by the wellhead at the seabed since that is where the casing hanger lands. The casing is usually landed using drillpipe from the hanger up to the rig floor. This is an easy and convenient way of landing casing on floaters, however, the use of drillpipe from the hanger to surface means that full size wiper plugs cannot be used as the ID of the casing will almost always be much greater than that of the drillpipe.

To facilitate pipe wiping in this case, wiper plugs are installed just below the hanger running tool. A ball dropped through the drillpipe will land on the lower plug, cause it to shear away from its carrier and allow it to wipe the casing.

Displacement

Cement should be displaced into place in a state of turbulent flow. Since displacement is usually done using rig pumps achieving turbulent flow is usually not a problem.

Preflushes, spacers and scavanger

Any mixing of mud and cement will reduce the quality of the cement job. Preflushes, spacers and scavengers are used to separate mud and cement and also to condition the hole to give better ultimate bonding between cement and formation.

Preflushes and spacers are usually made up of a combination of cement, mix water, weighting material and additives designed to help remove wall cake to give better cement/formation bonding characteristics. The weighting material is to provide primary well control.

Scavengers tend to be a diluted form of the cement slurry used in the main job. Adding cement obviously brings up the weight of the mix and if primary well control is a potential problem then the scavenger weight can be controlled by the amount of cement used.

The volume of preflush, spacers or scavenger used will depend on the perceived needs of the well. These should be sufficient to clean the hole and separate the mud and cement, but not so much that the hole will collapse.

SECTION 7: DEVIATION PROGRAMMME

The first thing to consider in planning deviated wells is whether they need to be deviated at all. Good operations management will ensure that Geologists and Reservoir Engineers are realistic about their subsurface target needs. If these needs can be met by a vertical well then they should be. Sometimes, however, it is necessary to drill a deviated well and, as programmers, we have to pick the best approach for the well. There are many options available today, all of which can be effective if done correctly.

To decide on the correct option for the well, the first consideration is how much drift from the surface location is required. High drift wells need to be kicked off high in the hole to achieve extended reach without having too high a hole angle. Small drift–wells should be kicked off deeper in the well due to the difficulty encountered in trying to hold hole direction with a hole angle of less than 15 °.

Hole angles in the range of 15 °–65 ° are relatively easy to live with, and with oil-based muds the upper limit can be considerably higher. Build-up rates are usually around 2½ °/100 ft or 30 m and this rate will give minimal operational problems, drillstring or casing wear. If possible, the 'build-up' section should be cased off, since an abrupt angle change anywhere in the well is a potential key seat area.

In shallow kick-offs jetting techniques using a Big Eye or Badger Bit can give kick-offs in soft formations quickly and cheaply. Abrupt angle changes are possible with this method and it is advisable to kick off with a smaller bit and then open hole prior to running casing (e.g., 12¼ inch Badger bit run followed by a 17½ inch hole opener prior to running 13⅜ inch casing). Directional drilling companies will try to dissuade you from using this method. Some people, less kind than the author, might suggest that their reluctance to use it stems from the minimal equipment requirements and the speed of kick-off, rather than any inherent weakness. In the right formation nothing will come close to jetting techniques for speed and cost-effectiveness.

The most common kick-off method in shallow kick-offs is the drilling motor mounted on a bent sub. This is oriented by using single shot surveys or telemetry and works quite effectively. Unless the set-up is totally steerable from surface though, the kick-off assembly must be pulled once initial kick-off has been achieved. This is time consuming and therefore costly.

For deep kick-offs the most common option is also the drilling motor mounted on a bent sub. Since the trip time from deeper kick-offs is longer than for shallow kick-offs, it can become cost-effective to use steerable systems in these cases.

Whipstocks provide a position deflection tool which can be used at any stage in the hole, but if they have to be used, the disadvantages of this tool should be borne in mind. The design weaknesses of the whipstock are the number of trips required to operate it and the abrupt angle change created. Abrupt angle changes can be tolerated in most wells, however, they can cause problems when running casing and they can erode drillstrings. Drillstring erosion is a function of the restoring force at the whipstock. The restoring force is a function in turn of the weight hanging below the abrupt angle change. In practice, this means that if you only have a few hundred feet or metres to drill to TD then you can probably live with a whipstock. If it is thousands of feet or metres then drillstring failure could prevent you from reaching TD and the tool should certainly be avoided.

SECTION 8: SURVEY REQUIREMENTS

Surveying is merely to determine exactly where the hole is. This is measured for two main reasons: reservoir management and relief well planning. In each country the energy authority will define minimum survey requirements. Clearly these must be met, however, it is possible that more stringent survey requirements might be needed by reservoir engineers, or for effective relief well planning. In order of increasing complication, the survey methods at our disposal are as follows:

Totcos are clockwork units which measure hole angle by using a pendulum mechanism with a pin on the end. They are crude but effective and since they only measure hole angle there is no need for monel drill collars in the BHA to overcome magnetic influences. A Totco ring is placed just above the bit inside the drill collars and the survey barrel is run in on sand line, slick line or dropped and recovered, Totco instruments are typically available in 0–8 ° units and 0–90 ° units and are unaffected by hole temperature.

Magnetic single shots have magnetic compasses built into them, in addition to the angle measuring units. Information is gathered using a small camera in the instrument, which takes a photograph of the compass heading and attitude. It is run in the same manner as the Totco, but does need monel drill collars to reduce magnetic influences. Furthermore it is sensitive to downhole temperature and unless 'heat shield' type units are used, the photographs or survey discs can 'burn up' before they can be recovered and developed.

Magnetic multishots are similar in principle to the magnetic single shot but with the additional ability to record magnetic inclination and direction on a film strip at regular depth intervals. A magnetic multishot barrel is dropped down to the BHA before it is tripped out of the hole. The BHA must have a non-magnetic DC installed.

Electronic multishots measure survey data using triaxial magnetometers which measure components of the earth's magnetic field and triaxial accelerometers which measure components of the force of gravity. These components are then interpretated to give inclination and direction. The housing for these sensors is identical to that of the magnetic multishots with the exception of the OD of the barrel being larger. They are used in a similar fashion, that is dropped into the BHA before pulling out of the hole.

Directional MWD in the form of a drill collar, is part of the BHA and measures survey data whilst drilling. The sensor (magnometer and accelerators) used in these tools are identical to the ones used in the electronic multishots. These are toughened and made suitable for the drilling environment. Such devices can be used with a high degree of accuracy, particularly when the data is computed at surface. This allows for the application of stringent calibration factors and temperature compensation.

The data from these tools is transmitted to the surface in real-time. Any one of three types of transmission system (positive or negative mud pulses, or the continuous wave system) can be used. This choice depends on the service companies.

SECTION 9: MUD PROGRAMME

The function of a mud system is to:

- Maintain primary well control
- Carry cuttings out of the hole and hold them in suspension when circulation stops
- Release cuttings at surface
- Clean the bit and the bottom of the hole
- Cool and lubricate the bit
- Transmit hydraulic power to the bit
- Minimise reaction between the drill fluid and formations drilled
- Provide support for open hole
- Coat the wall of permeable formations with a thin impermeable filter cake
- Lubricate the drill string/open hole interface
- Allow the transmission of information about the hole being drilled

In practice, mud programming can be broken down as follows:

1. Determination of mud weight requirement to maintain primary well control.
2. Determination of suitable 'trip margin' which is added to the primary well control weight to give a programmed mud weight.
3. Confirmation that this mud weight does not exceed formation strengths when considered in a dynamic (i.e., circulated) mode.

4. Analysis of formations to be drilled and the likely reaction of these to the available drilling fluids alternatives. Using this information to select a basic mud type such as:

Water-based	freshwater muds seawater muds calcium muds lignosulphonate muds polymer muds
Oil-based	invert oil emulsion muds environmentally sensitive oil-based mud true oil mud

5. Determination of fluid loss requirements.
6. Determination of pH requirements
7. Determination of viscosity requirements.
8. Determination of temperature stability requirements.
9. Analysis of rig mud treatment equipment to meet hole requirements with selected mud types.

Water-based muds are the most commonly used types of mud. Onshore, these are normally based on freshwater with the addition of barites, clay, polymers or other chemicals to provide the designed mud properties. Since most formations were laid down in a marine environment, it is quite common for formations such as shales to react to 'straight' freshwater muds. If this is anticipated, an inhibited system can be created by adding gypsum, lime or potassium chloride. If large sections of salt are anticipated, then the mud system must be inhibited by supersaturating with salt.

Offshore, saltwater-based muds are used as there is an abundant supply of the raw material. Special clays are used to provide acceptable yields when used with salt rather than freshwater.

Dispersed systems are effective when drilling though high-activity clays. These use thinners or dispersants to make the clay platelets separate in the water phase and provide better viscosity and filtration control. Bentonite–Spersene (a chrome lignosulphonate) is used a lot, however, the environmental implications of using chrome lignosulphates is promoting the use of chrome-free lignosulphonates such as calcium lignosulphonate.

Viscosity is controlled in water based fluids by the addition of clays (bentonite) and polymers (CMC–Carboxy Methyl Cellulose). CMC also has the benefit of reducing fluid loss from the mud in permeable formations and building up a filter cake on the wall of the hole. To combat hole problems, sometimes a thin rubbery wall cake is required and to achieve this lignite–caustic soda based muds are utilised. Differential sticking (covered in chapter 12) can be a function of the depth of wall cake, so a compromise must be reached between effective fluid loss and the creation of potential stuck pipe problems.

Water based muds should be kept alkaline. This is done by the addition of caustic soda in most cases.

Oil-based muds have evolved from the original Black Magic Muds, through the invert oil emulsion muds of the 1970s, to the present range of oil-based muds which have a reduced environmental impact. In some wells, notably those with shale problems, oil-based muds have reduced, and, in some cases, eliminated hole problems. They create a relatively inert drilling fluid which does not have an easily accessible water phase to react with shales and cause hydration.

The water phase of most muds is in the range of 10 – 30 per cent and is typically kept more saline than anticipated formations by adding calcium chloride. Emulsifiers are used to enclose the water droplets in the mud. The water phase is used to support barite, contribute to gel strengths, act as a fluid loss agent and allow chemical additives to dissolve in the mud. The stability of the oil/water emulsion can be measured by recording the mud's electrical stability. An additional check can be made by carrying out a HPHT fluid loss test and checking the filtrate for water content.

As with water-based muds, special chemicals are available to treat oil-based muds and control viscosity and fluid loss. Typically, these are not the same chemicals as used with water-based muds.

Oil muds are valuable in the right application, however, it must be realised that they do have disadvantages too. They are dirty and this creates problems for the rig crew, the rig and the environment. They can react with water so all valves in the mud system had better seal if you want to retain the original properties. They can absorb gas at the bottom of the hole into solution and release it suddenly as the gaseated mud comes to surface. They create cuttings disposal problems as the cuttings cannot be just tipped into the sea as was our wont with other cleaner sytems. They are expensive and usually need a mud engineer to maintain them. Finally they can create problems in the shaker house if there is insufficient ventilation to clear the fumes and vapours created therein.

In addition to the usual mud additives used to maintain weight, viscosity and fluid loss special mud additives are used to cope with special situations such as:

1. Lost circulation material (LCM)
2. Stuck pipe freeing fluids
3. Lubricants/detergents
4. Corrosion control agents

LCM Lost circulation is covered in detail in chapter 12. Primary causes of total losses are fractures and cavenous formations. Porous and permeable or unconsolidated zones cause partial lost circulation or seepages and in time can actually 'heal' themselves. The function of lost circulation material is to plug the holes or porosity to prevent the passage of fluid into the formation. They are basically mechanical barriers. Since porosity varies, then the size of barrier or LCM that we require also varies. All of the most commonly used LCMs, i.e., mica, nut plug and cellophane come in three sizes: course, medium and fine. If the exact source of the lost circulation problem is not known then a mix of all the sizes should be used.

Stuck pipe freeing fluids In the event that pipe becomes stuck in the hole, various stuck pipe freeing fluids are available to help free it by chemically attacking the filter cake at the point of sticking.

Lubricants/detergents Bit and stabiliser balling can be a problem when drilling through some formations. If this is severe, swabbing can occur leading to a risk of well control problems. In water-based systems, detergents can be added to reduce the surface tension of the water and help reduce balling. If trips out of the hole are dirty (i.e., stabilisers packed off with clays), the addition of a lubricant or detergent can reduce or eliminate this effect. In sticky holes, where high torque is required just to rotate the drillstring, they can provide a means of getting more torque to the bit without exceeding the drillpipe make up torque which is the usual limiting factor in most cases.

Corrosion control agents Some formations produce corrosive agents such as carbon dioxide and hydrogen sulphide. Inhibitors or scavengers are used to combat this problem.

Having decided on the mud system required for a well, the mud treatment equipment available on the rig should be appraised to check compatibility with the selected system. The treatment equipment falls into four main groups:

Treatment equipment	Solids removed in microns
Shale shakers	60+
De-sanders and De-silters	60−15+
Mud cleaners	30+
Centrifuge	1−2+

Shale shakers These are the single most important component in the system since they are the initial cuttings removal system and remove the major portion of drilled solids circulated out of the well. Double-deck shakers are fairly standard these days and they allow a course upper and fine lower screen to increase cutting removal efficiency.

Mud cleaners By employing both de-silting hydrocyclones and fine mesh vibrating screens, mud cleaners effectively remove fine drilled solids from weighted muds without excessive barite loss.

De-sanders and de-silters Employ hydrocyclones to remove drilled solids. The smaller the hydrocyclone, the finer the solids that can be removed will be.

Centrifuge The last stage in the treatment process is usually the decanting centrifuge which can remove solids down to 1 or 2 microns in size. They are particularly important when using oil-based muds. Barites can be recovered for reuse in the mud system.

Having specified a mud type and fluid parameters, thought must be given to the ability to maintain these designed parameters at the well site. In order to check parameters, the following tests need to be carried out. Dependent on the nature of the well, the frequency of testing should be determined. Typically, the mud will be completely tested twice a shift. However, its weight will be monitored continuously to maintain primary control. The Drilling Programme must specify the test frequency.

Weight measurement This is done with a simple mud balance which can be calibrated by using freshwater. This should indicate a weight of 8.33 ppg at 70 °F.

Viscosity measurement This is done with a marsh funnel. This is a conical funnel with a fixed diameter outlet. It is filled with mud to a predetermined level and then the time taken for 1 U.S. quart (946cc) to run out is measured. This too can be checked using freshwater which should indicate a reading of 26 seconds.

Gel strength measurement This is done with a viscometer. Viscometers work using two concentric cylinders. The outer cylinder is rotated and friction in the mud between the two cylinders applies torque to the inner cylinder which can be measured. From this, plastic viscosity and gel strength can be calculated. Most viscometers run at two speeds 300 RPM and 600 RPM.

Filtration This is measured using a filter press either at room temperature or with high pressure/high temperature equipment. The HPHT equipment is used to simulate more closely the downhole conditions of the mud. How true this is, is open to debate, as the mud downhole is in movement for most of the time and therefore dynamic, whereas the filter press is of course a static measurement.

The filter press operates by applying pressure to a given volume of mud. The only way that pressure can be relieved is through a layer of filter paper in the base of the apparatus. Fluid is forced through the filter paper leaving a filter cake on the filter paper. The amount of fluid measured over 30 minutes in cubic centimetres is termed the API fluid loss. The conventional filter press operates at 100 p.s.i. whilst the HPHT unit uses 500 p.s.i.

Sand content This is measured using a plastic cylinder with 200 mesh screen, a funnel that fits on to it and a glass measuring tube. A predetermined amount of mud is poured into the cylinder. Water is added to the mud in the case of water-based mud, or diesel in the case of oil-based mud and the cylinder agitated. The mixture is then poured onto the 200 mesh screen which separates out the sand. The cylinder is washed with more fluid (water or diesel) and the mixture again poured on the 200 mesh screen. The sand left on the screen is washed into the glass measuring tube where it is allowed to settle. The volume of sand is read off the graduations in the measuring tube and quoted as percentage sand in the mud by volume.

Liquid and solids content of mud This is measured using a solids retort kit. This takes a given volume of mud and using heat distils off the liquid phase. This liquid phase is then condensed and measured. The solids percentage is merely calculated by subtracting the liquid percentage from 100.

pH This is a convenient means of measuring the alkalinity of the mud. Distilled water which is considered to have neither acidity nor alkalinity is therefore neutral and has a pH or 7. Higher pH values represent higher alkalinity. Muds are typically kept in the 9 – 10.5 range. This is mainly to keep drillpipe and casing corrosion to a minimum, to restrict potential formation damage and to allow some of the added chemicals to function better.

pH is checked by using pH papers whose colours change upon immersion in a fluid indicates the pH of that fluid, or alternatively by using a pH meter which measures electric potential generated inside the meters electrode when immersed in the mud.

SECTION 10: BIT AND HYDRAULICS PROGRAMME

Bit selection and hydraulic optimisation is covered in detail in chapter 11. The fundamentals for drilling optimisation are as follows:

Bit selection To select a bit the following factors are considered:

formation, drillability and characteristics
mud system in use
directional implications
bit drive methods
availability

Formation drillability A given formation can be reasonably expected to have a given drillability. The best indication of this is bit performance in nearby wells. If this information is available then selection is made easy. On wildcats, however, some assumptions have to be made and then the results of actual runs analysed.

Tricone bits for soft formations drill by a shearing and twisting action. They gouge their way into the formation and sometimes can drill so fast that the penetration rate must be controlled to allow efficient hole cleaning. In soft to medium-soft formations that are not too sticky, PDC bits offer the best alternative, especially in oil-based muds.

Selecting exactly the right bit requires field experience, however, it will usually be worth the effort of trial and error to determine the correct PDC bit especially for development drilling.

Harder formations will be drilled by using insert bits with journal bearings and gauge protection or alternatively diamond bits. Journal bearing insert Tricone bits can stay in the hole for several days so they must always be used with shrouded nozzles to prevent washing out around the circlips which retain the nozzles.

Mud systems in use If oil-based mud is being used, it is probably to control shale problems. In this case, PDC bits will probably offer the best choice of bit. PDC bits do also work effectively in water-based mud but not as well as they do in oil-based muds.

Sealed bearing Tricone bits should be used, if the drilling fluid will contain abrasive particles, to prevent premature bearing failure.

Directional implications Most Tricone bits used on rotary will exhibit some right-hand walk tendency. PDCs on the other hand due to their symmetrical cutting action tend to drill straight ahead. If the well is planned to be drilled with Tricones, then it is advisable to lead the well to the left and allow the natural tendency of the bits to pull the well back on course. If, however, during the drilling of the well, direction must be held or a correction run made, it makes sense to run a PDC bit in preference to a Tricone to obviate the need or potential need for the correction run.

Bit drive methods Bits are driven either by the rig's rotary table (or top drive system) or by a downhole motor. The advantage of using a downhole motor is that torque is applied directly at the bit and the amount of wear on the drillstring and casing is reduced. The disadvanges are the relatively high rental cost of the motor, its limited operational life and, in some cases, the high rotary speed at which it rotates the bit. In most cases, drilling turbines will turn too fast for reasonable bit life on a Tricone bit. This restricts their use to diamond or PDC bits. New generation drilling motors working on the positive displacement principal, on the other hand, can be designed to rotate slowly enough to allow the use of tricone bits. Turbines do exhibit a tendency to walk to the left whilst drilling and this is a feature which can be used to advantage when drilling a directional well.

It is worth remarking on the programme that should PDC bits and/or diamond bits be programmed, then a junk sub should be run in the string in the previous bit run. This is to prevent small items of junk damaging the PDC or diamond bit. It should also be realised that these bits have a different cutting profile to that of a flat bottomed Tricone bit. They should therefore be bedded in gently before drilling ahead with them.

Once the mud has been programmed, the hydraulics implications of using this mud should be considered. 'Hydraulics' in this case, just means looking at the mud dynamically rather than statically. In section 7, we listed the desirable functions of mud. Some of these can be considered to be static functions, such as primary well control, whilst the others will dcpend to a great extent on the dynamics, i.e. how much mud is being pumped and at what pressure. The main mud functions effected by dynamics or 'hydraulics' are to:

transport cuttings out of the hole, a function of annular velocity
clean the bit and the bottom of the hole
cool and lubricate the bit
transport hydraulic horsepower to the bit
minimise reaction between drill fluid and formations, a function of AV

Annular velocity (AV) A certain minimum annular velocity is required for a given mud type to prevent cuttings slippage through the mud and to effectively lift the cuttings out of the hole to the mud treatment equipment. The most commonly quoted figure used to be 100 ft/minute in the annulus. In fact, most muds will lift cuttings quite effectively at AVs below this figure and each system should be studied individually to determine AV requirements. The heavier the mud in use, then the slower the acceptable AV will be. Taking this trend to its extreme, when the mud gets very heavy, the cuttings will float to the surface. This point is critical when drilling high pressure wells where the equivalent circulation density of the mud being circulated can be enough to induce losses downhole if 100 ft/minute AV were to be used. Having decided upon the required AV, the volume to be pumped can be calculated.

Bit and hole cleaning Drilling fluids flow in either a laminar or turbulent manner (or in a combination of these). Laminar flow will cause less hole erosion, however, turbulent flow is likely to clean the hole better. In practice, laminar flow and turbulent flow are acceptable in the annulus. Critical mud velocity in the annulus which occurs when flow changes from laminar to turbulent, or vice versa, should always be avoided however, as it is during this flow that the maximum erosion of the well bore will occur.

There are two main theories concerning how much hydraulic horsepower should be expended at the bit to gain maximum cleaning efficiency. The first theory is the maximum hydraulics horsepower theory, which in practice means expending ±65 per cent of the available HHP at the bit. The alternative theory is the maximum jet impact theory which in practice means expending ±48 per cent of the available HHP at the bit.

At the time of writing, the author still has not worked out which of these theories is the best one to use for a given set of circumstances. There are practical limitations to the minimum size of jets to be used in a bit. (I would regard 8/32 inch as an absolute minimum to prevent plugging). This fact, coupled with the length of bit runs being long enough to affect the parasitic friction losses in the system, means that a sense of reality must prevail when selecting bit nozzles.

If the bit is performing effectively and the hole is being cleaned, then the hydraulics are adequate. In practice, this will mean that the HHP being expended at the bit is probably in the range of 40 – 70 per cent. Innovative techniques and equipment such as cavijet nozzles from Smith International and extended nozzles do have benefits but their use is still not very common.

Usually the main limitations to the hydraulics of a system is what the rig pumps can reliably deliver. Having gone through a 'calculation period' in the 1970s, the author has adopted a more practical approach to hydraulics which seems to be just as effective. All it entails is:

1. Work out AV, requirements to clean hole.
2. Convert this into flow rate.
3. Work out parasitic losses in the drill string at this flow rate at the anticipated end of the bit run using a hydraulic programme or hydraulic slide rule.
4. Find out what pressure the rig pumps can deliver at this flow rate.
5. Using a hydraulics programme or a hydraulic slide rule, calculate the nozzle requirements.

Note that the author still advocates the use of the hydraulic slide rule. Their use has been derided in the industry since the introduction of programmable calculators and PCs. It is true to say that they do not calculate pressure losses as accurately as a computer could. If, however, the accuracy of the measurement of weight, PV, viscosity etc., is analysed, coupled with the fact that these measurements are made statically at room temperature when the mud's performance must be estimated dynamically, then it can be seen that realistically the hydraulic slide rule is as accurate a tool as is needed. Most industries in the world would love to have tools like our hydraulic slide rules and field table books and the author finds it lamentable that the drilling industry is rejecting these on totally fallacious grounds of accuracy.

SECTION 11: EVALUATION REQUIREMENTS

In this part of the programme, the evaluation requirements necessary to meet the well objectives (see also chapter 8) should be formatted as follows:

Drilling log requirements:
Mud logging requirements:
Coring requirements:
MWD requirements:
Electric logging requirements:
Testing requirements:

SECTION 12: OPERATIONAL PROCEDURES AND TIME DEPTH GRAPH CONSTRUCTION

The degree of detail required in the operational procedures part of the Drilling Programme is dependent on how comprehensive the Drilling Operations Manual is. If it is comprehensively written, then no operational procedures should be included in the Drilling Programme. If there is no manual available or its quality is poor, then the operational procedure section of the Drilling Programme should be a step-by-step guide to what has to be done in the well.

The Time Depth Graph is a tool used to show the expected well status at any time from spud to completion of plug back and abandon. To construct a Time Depth Graph certain assumptions and estimates must be made. If there have been other similar wells drilled in the area then a Time Depth Graph can be based on their performance with few assumptions required. When drilling rank wildcats the assumptions will outnumber the available data.

The phases of most exploration wells can be listed along the following lines:

Phase	Time
Pre-spud preparation	
Drilling top hole	
Run and set surface casing	
Drill	
Set intermediate casing	
Drill	
Core	
Drill	cont.

Phase	Time
Log	
Test	
Plug back and abandon	

The Time Depth Graph uses time along the x-axis usually in days and depth along the y-axis with zero depth at the upper end of the axis. Both axes should be approximately 20 per cent longer than initial time and depth estimates would indicate to allow programme changes.

SECTION 13: SITE PLAN

A site plan should be included in each programme showing the following features:

For all wells: The licence area in which the well is being drilled. The position of the well relative to other outstep wells.
For land wells: Access roads and instructions on how to reach the site.
For offshore wells: Details of any sea bed obstructions, pipelines etc, in the area that the well is being drilled in.

In practice, to satisfy the above requirements, two plans will need to be included in most Drilling Programmes. Some operators like to include a reservoir map in this section also.

SECTION 14: REPORTING REQUIREMENTS AND CONTACT NUMBERS

In this section the reporting requirements are listed. This should lay down clearly who should receive what information or samples, by what means, when and how often. All relevant phone numbers and addresses must be included in this section.

Chapter 8

EVALUATION

If we are not drilling the well to produce oil and gas or to aid production of oil, then we are doing it to gain information. This gaining of information is termed evaluation. We have many means at our disposal to determine what we are drilling through and more importantly, whether these formations could contain commercially extractable hydrocarbons. These means can be grouped under six headings:

1. Drilling log
2. Mud logging
3. Coring
4. MWD
5. Electric logging
6. Testing

Each of these evaluation means is discussed separately within this chapter. Each one supports the others and by utilising some or even all of them, a true picture of the well's commercial potential can be gained.

With the exception of the drilling log (which is a fairly passively recorded tool), they all cost money, in some cases a lot of money. When we engineer wells, we must ensure that the evaluation requirements of the wells are met in the most cost-effective manner.

In chapter 5, we looked at a typical North Sea exploration well. Out of the US$ 5 300 000 overall well cost, US$ 750 000 was spent directly on evaluation. If we add in rig time as a cost, then over $1 000 000 is being spent directly on evaluation in this type of well. This represents around 20 per cent of the well cost and as we discussed in chapter 5, this is a large enough percentage of the well cost to mean that even a small saving is worth pursuing.

1. DRILLING LOG

This is a very simple record of penetration rate against time. It is easily plotted manually or on multipen recorders such as the Rig Serv or Totco units. The principle behind this is simply that different formations have different drilliability for a given set of drilling parameters. Changes in formation can be noticed and recorded. These changes can be correlated with nearby well records, seismic records, cuttings or known local geology. It is an unusual well that could be completed on the basis of a drilling log alone. However, as a correlating tool it is cheap (or free) and very useful.

2. MUD LOGGING

Mud logging is probably the most under-utilised form of evaluation service in use today. Most rigs have Mud Loggers on board but in the author's opinion, use them to about 10 per cent of their capability. A modern mud logging unit, staffed with experienced graduate geologists, is a 'real time' database for the operator to use. They record every parameter which indicates formation, drillability, hydrocarbon presence and pressure trends. Despite this, they are still treated as the poor relation to the Electric Logger whose role, in fact, is just to confirm that what the Mud Loggers suggested was there, was really there.

Mud logging units have benefited from the introduction of the silicon chip and these days most are fully computerised. This has meant an increase in the services that they offer and a reduction in the weight of their unit. This latter fact is significant when it comes to considering topside loading on platforms or deck load on floaters and jack-ups.

Early mud logging units simply offered a sample catching and gas detection service. Modern units still perform these functions but offer more facilities. The modern unit will provide the following:

1. Sample catching and description service
2. Gas detection
3. Gas analysis
4. H_2S detection
5. Oil show detection and analysis
6. Pit level monitoring
7. Trip tank monitoring
8. Drilling parameter monitoring
9. Penetration rate monitoring
10. D exponent read-out
11. Pressure prediction
12. Mud property monitoring
13. Flowline temperature monitoring
14. Cuttings bulk density measurement

2.1 Mud logging services

These services are explained in detail below.

2.1.1 Sample catching and description service

The service contract that the operator has given the mud logging company will stipulate catching formation samples at predetermined intervals. These intervals will be selected to meet the Reservoir Engineer/Petroleum Geologists' needs and also to comply with any local energy authority requirements.

The mud logger will catch samples off the shale shaker having calculated 'bottoms up' time from the hole geometry and circulation rate. In addition, he will also take samples from the underflow of the de-sander and de-silter. A good shale shaker will take out 85 per cent of drilled solids, however, fine particles can pass through the shaker screen, and in some cases could pass unnoticed if they were not caught leaving the de-sander and de-silter.

Some of the tests carried out on the cuttings are described in detail later in this section. In addition to these tests, the Mud Logger also describes the formation samples caught as follows:

Formation	type, hardness, cementation, colour, prosity, lustre.
Grain	size, shape.

2.1.2 Gas detection

Mud Loggers detect gas in two forms. Firstly, gas in the mud is detected in a gas trap fitted in the flowline. This is simply a steel box immersed in the flowline mud flow which is fitted with a paddle agitator and a pipe which, by using negative pressure, can draw in any gas released by the agitation, to be detected in the unit. Secondly, gas is detected in the cuttings caught as described above. If gas is present in the formation being drilled and the permeability of this formation is poor, then by blending and crushing the samples, gas can be released and detected.

The relationship between 'free' gas in the mud and 'trapped' gas in the rock allows the Mud Logger to estimate downhole conditions of porosity and permeability.

Constant monitoring of gas levels in a well is an excellent indicator for the Drilling Supervisor to use to assess what is going on downhole. An experienced Mud Logger is worth his/her weight in gold when it comes to determining gas levels attributable to the amount of hole drilled and excessive gas levels, which indicate a primary hydrostatic underbalance. As the levels are continuously plotted on the mud log the Drilling Supervisor can see the effects of connection gas. (When mud is being circulated whilst drilling, the friction of the mud column in the annulus creates a higher bottom hole pressure than when circulation has stopped. This fact, plus the swabbing effect of pulling the kelly back to make a connection, means that the gas levels corresponding to 'bottoms up' circulation from a connection will quite often be higher than the average 'background' gas levels. Background gas levels can be used to estimate formation pressure trends. As formation pressure increases, then a rise in background gas levels can be expected, if the formations being drilled are gas bearing.

'Trip' gas manifests itself in the form of higher gas levels once 'bottoms up' is circulated after a round trip. The theory behind this is similar to connection gas (reduction in bottomhole pressure when the pumps are off, allowing more gas to enter the borehole from the formation).

Connection gas, background gas and trip gas viewed in isolation do not mean very much. They should be considered as trends to be observed and related to formation samples returned and potential downhole pressure trends.

The gas detection equipment installed in most mud logging units is the Chromatograph. This is a device which has a filament heated by the passage of electric current (hence the expression hot wire chromatograph). As gas is passed over the hot wire, it combusts, heating up the filament further and increasing its resistance. This change in resistance is measurable and by calibrating the Chromatograph with a known gas, reasonable accuracy can be obtained. This is commonly done by putting 100 ml of calcium carbide into the drillstring at a connection. When circulated around the hole it creates a known volume of acetylene which comes back to surface to be 'detected'. Gas levels are plotted on the mud log.

2.1.3 Gas analysis

The Chromatograph will give an indication of the make up of the gas being tested by subdividing it into:

methane
ethane
propane
butane
pentanes

The device does have limits to the percentages of each component that it can handle so, in some cases, it is necessary to dilute the gas sample with air prior to testing. If it is vital to know the chemical composition of the gas accurately, then it is worth stipulating flame ionisation gas detection systems. For most wells, though, the Chromatograph will meet evaluation needs. The Chromatographic breakdown of detected gas is plotted on the mud log.

2.1.4 H_2S detection

Although most rigs have independent H_2S detection equipment, the Mud Logger provides a useful backup system. The Mud Loggers use a system for H_2S detection which relies on the adsorption properties of H_2S. As H_2S passes over a semi-conductor within the H_2S detection instrument, it is adsorbed by the semi-conductor. This changes the resistance of the semi-conductor and the change in resistance can be measured.

2.1.5 Oil show detection and analysis

Mud Loggers detect oil shows at two stages. Firstly, in the microscope, and secondly in the ultra violet light box. Under the microscope, some cuttings can be readily seen to be oil stained. By placing cuttings under an ultra violet light in a UV light box, oil shows can be seen between fluorescence caused by external sources such as pipe dope and that caused by hydrocarbons from downhole. From the appearance and colour of the fluorescence, an estimate of the grade of oil present can be made. Furthermore, by observing how the fluorescence is distributed throughout the grains, an estimate of whether the formation has oil in fractures or within porosity can be made.

2.1.6 Pit level monitoring

The Mud Logger provides a backup pit level monitoring system to complement the pit volume totaliser system found on most rigs. Good pit level monitoring is vital in the early detection of well kicks or losses and the Mud Loggers plot pit levels on a continuous basis.

2.1.7 Trip tank monitoring

As with pit levels, the Mud Logger provides a backup service for the rig's own equipment. Furthermore, as with the pit levels, monitoring this service must only be considered as a backup and not a replacement for the Driller's vigilance and attention to his own instrument.

2.1.8 Drilling parameter monitoring

The mud logging unit can record the following drilling parameters:

hook load
pump pressure
mud flow in
mud flow out
pump speed
mud weight in
mud weight out
mud temperature in
mud temperature out
mud resistivity in
mud resistivity out
rotary torque
rotary speed
shut-in casing pressure

Depending on the logging unit, up to 12 parameters can be continuously plotted on a read-out and on the ultimate mud log.

2.1.9 Penetration rate monitoring

This is recorded either manually or passively and provides a backup for the Driller's own records.

2.1.10 D exponent read-out

Jordan and Shirley developed an equation which created an indicator of pressure trends called the D exponent. The basic D exponent has been modified to include a mud weight component to give the DC exponent. It should be realised that any D exponent should be used as a trend indicator only and not in absolute terms. Reading too much into a D exponent in isolation can be misleading. The trend in D exponents, however, do indicate changes downhole and we are interested in these. If the D exponent is increasing slowly as we get deeper in the well and then suddenly increases rapidly with depth, this indicates a change downhole. This may or may not be significant, however, providing that we are aware of it, we can check it out. The Mud Loggers print out the D exponent on a continuous basis and this is plotted on the mud log.

2.1.11 Pressure prediction

A continuous pressure prediction level is calculated from drilling performance and this is plotted on the mud log.

2.1.12 Mud property monitoring

The modern mud logging unit will display and record mud weight and resistivity.

2.1.13 Flowline temperature monitoring

Using detectors in the flowline, the temperature of the mud returns in the flowline can be measured and recorded. As with many parameters recorded by the Mud Logger, the flowline temperature does not mean very much when looked at in isolation. What is more relevant is the trend of flowline temperature. As the well gets deeper, then the flowline temperature will increase gradually. Any change in the rate of temperature increase is cause for concern and should be investigated prior to drilling ahead.

2.1.14 Cutting bulk density measurement

Cuttings, especially shale cuttings, can indicate their degree of compaction by virtue of their bulk density. Their compaction in turn will indicate possible downhole pressure regimes. By using a flask filled with a variable density liquid, the Mud Logger can measure the bulk density of the cuttings. The cuttings are dropped into the flask and their final settling depth indicates their bulk density.

3. CORING

In petroleum drilling, geologists are asked to determine models of reservoirs. This is to determine both the size of the reservoir, as well as how much hydrocarbons the reservoir contains. Part of this modelling requires determining how various fluids will flow through the reservoir. The economics of development and the manner of development should reflect the nature of the reservoir and mistakes in the accepted reservoir model can be very costly, out weighing the cost of the odd well or two by a considerable factor.

There are therefore, three reasons for coring a reservoir, all of which are required to derive the reservoir model:

1. To obtain sedimentation detail. Sedimentation detail means being able to measure the thickness of the sequences of sandstones, shales, evaporites etc. Also

important are the features within these layers such as, bedding planes and faults. This can be accomplished with 'standard' coring techniques.

The details can be looked at in more depth if add-on techniques are used. These techniques do not have universal application but will be requested by the more enlightened geologists on a selective basis. The techniques include orientated coring, sponge coring and pressure coring.

2. To obtain reservoir engineering detail. For a given reservoir exploitation, it is important to know what fluids are contained within the reservoir and how these reservoir fluids behave. Are they waxy, are they water bearing etc.? It is important to know how fluids will flow in the various directions within the reservoirs and whether certain fluids will damage the reservoir structure, making it more difficult to get hydrocarbons out as time passes (for example, the reservoir may have shales that will be liable to swelling if water is injected to enhance recovery or if oil/water contact point is allowed to rise). Again, this can be accomplished using standard coring techniques but better results may be needed that will require the add-on techniques.

3. To validate/calibrate electric logs. Electric logs depend upon a base measurement from which interpretation is made to decide the trigger levels that identify formation changes. These trigger levels are not arbitrary but to get accurate results, the only way to validate the measurements is to compare the logs to a piece of rock.

3.1 Coring technology

Coring is the technique of cutting a cylinder of rock and bringing it to the surface where it can be examined. The types of core barrel are many but fall into two main categories.

1. Conventional, where the entire core barrel is withdrawn to surface to recover core.
2. Wireline, where only the inner tube is brought to surface to recover the core. This is done by means of a wireline.

In the petroleum industry, wireline systems are rarely utilised. This is predominately because the drillstrings used are too small to allow the passage of a core size adequate for the engineering data to be determined.

The coring operations falls into three phases:

1. Cutting the annulus
2. Getting the core into the core barrel
3. Holding the core in the core barrel for as long as required.

3.1.1 Cutting the annulus

It is important to consider the system used to cut the annulus. This means being selective about the drilling fluid and core barrel as well as about the cutting head. The techniques used are the same as those employed for normal drilling and it should be noted that most coring systems employed in the oil industry cut 20 per cent less material than drill bits in a given hole size. To optimise penetration rates, coring systems should use 80 per cent of the flow rates and power that would be used for full drill bits. In certain circumstances, this may need to be waived as the objective of recovering the core may conflict with achieving maximum ROP.

It cannot be over stressed that drilling fluid properties have a major impact on the success or failure of a core job. **Mud weight** should be such that the over balance is at least 100 p.s.i. and for soft formations 300 to 400 p.s.i. minimum. The **fluid viscosity** should be controlled in such a way the risk of swabbing is minimal. In particular gel strengths should be low. In order to prevent fluid invasion as much as possible, the standard **fluid loss** should be kept below 10 ml.

Coreheads can use the same cutting structures as drill bits that is:

polycrystalline diamond compacts (PDCs, Stratapaxs)
thermally stable polycrystalline elements (TSDs or TSPs)
natural diamonds
roller cone cutters

A mixture of these cutter types can also be employed. As with drill bits, the type of cutting element is determined by the formation to be cut. They all have valid successful applications and selection should be based on expected lithology and previous well records. Coreheads are used in a different way to drill bits and the design requirements are not the same. Whereas a drill bit should be designed for both ROP and longevity, a corehead run is limited to the length of barrel used. Although this is increasing, most core runs will not exceed 90 ft.

As with any cutting mechancism, core barrels must be supplied with an adequate fluid supply to clean and cool the corehead. Cutting performance will be determined also by the power that is transmitted to the coreheads. In recent years, the amount of fluid and power in the form of torque has been steadily increasing and the core barrels should be designed to handle this increase. Core barrels have changed little in the past 35 years and many designers are woefully remiss, allowing operators to select the operating parameters they desire.

A fairly recent study showed that 30 per cent of equipment shipped to rigs were returned with outer tube connection damage and as a result there has been development in the North Sea of outer tubes capable of resisting higher torque.

Nearly all core barrels restrict internal flow rates quite dramatically. This is due to very low internal clearances. The effect of this is to keep ROPs to a low ceiling.

3.1.2 Getting the core into the core barrel

Originally core barrels were single tube items and the advent of a double tube system, where the inner barrel was independent of the outer rotation, improved core run lengths and recovery rates dramatically.

In bearing assemblies, it is generally accepted that the lower the torsional connection between inner and outer the better. Most core barrels have the inner tube hanging from a simple roller ball thrust bearing located in the top of the core barrel. The bottom of the inner tube is kept central in the corehead by virtue of reduced clearances in a fluted area. Some core barrels have additional stabilisation along the inner tube length. Some core barrels have a more complicated bearing system which may give advantages. Generally, the simpler the system the less is likely to go wrong, provided of course, it is designed to cope adequately with the loads encountered.

The advent of a two tube system gives rise to the problems of spacing. It is important that the inner tube assembly is close to, but not touching, the outer assembly at the corehead. If it is too high, the core may be exposed to mechanical and fluid forces that are undesirable. If it is too low, then the inner tube will no longer be free and will turn. As the inner tube is in contact with the core by virtue of the core catching mechanism, it will tend to try and turn the core. This will result in under gauge core or breakage of the core at zones of weakness and can result in core loss.

It is important that the outer tube runs reasonably true and does not bend under normal coring weight. Unfortunately, there are very few core barrels that meet this, as the requirement of maximum core size and external fishing dimensions mean that it is difficult to achieve both requirements fully. Indeed, some core barrel outer tubes are thought to deflect so much that they can come into contact with the inner.

There are four inner tube systems offered:

conventional steel inner tubes
expendable GRP (glass fibre reinforced plastic) tubing
expendable aluminium tubing
inner sleeve tubing (triple tube system)

The last three systems are attempts to both improve the handling of the core, and also to reduce the friction coefficient between the core and inner tube. Certainly, there are substantial benefits to be achieved in terms of core quality by using expendable tubing. The core delivered to core analysis laboratories receives much less damage by virtue of being handled less. There is some evidence that core recovery and the incidence of core jamming are improved when using expendable tubing. The friction coefficient is lowest for GRP, followed closed by aluminium. Both suffer from thermal expansion problems which cause difficulty in establishing the correct spacing out of the inner in the outer. This is primarily because accurate figures for downhole operating temperatures are not normally available. The use of these expendable tubes has received increasing acceptable of late. This is primarily due to the improvements in ancillary equipment that makes the well site use safe and slick.

3.1.3 Holding the core in the core barrel for as long as required

Many varieties of core catchers are offered. The standard core catcher is a split spring housed on a tapered shoe. As the relative motion of the core occurs, the spring slides down the taper and closes, thereby reducing its ID and tightening on the core. It is too large to pass through the bottom of the housing and the continued upward movement starts to pull on the core which eventually breaks. A variation of this is the slip catcher. This operates on the same principle, but instead of a spring there are three or four small wedges that slide in their own groove. Both of these catchers work well in all formations including unconsolidated formations, although care must be taken on the tip out of hole. Also offered are dog catchers, finger baskets and flapper catchers. These are aimed at soft formations but can create more problems than they solve. Soft formation coring is a special problem which requires special care but can be successfully conducted using virtually standard equipment.

3.2 Job planning

The following list is normally requested by the coring service company to evaluate the equipment required:

1. Hole size to be cored in
2. Minimum acceptable core size
3. Mud system anticipated
4. Directional programme
5. Target formation depths
6. Target formation lithology
7. Expected method of core point identification
8. Anticipated coring interval
9. Special coring techniques, e.g. core orientation
10. Rig limitations – derrick height, top drive or rotary table etc.
11. Offset data – preferably core logs but bit records will do

From this information, a detailed programme can be established selecting equipment to meet the job requirements. If the well is highly exploratory, then it is normal to take standard coring equipment with a selection of coreheads to allow for contingencies. If there is a reasonable amount of reliable offset data then a pre-job analysis can be conducted which will give the coring operator an idea of expected performance. This is often invaluable as the Service Engineer generally 'feels' his way, modifying equipment selection after each run. This can be a costly exercise if his expectations do not meet quite closely to reality. However, the nature of most wells is variable, which is why the attendance of an experienced coring operator is invaluable.

When planning coring operations in a well, the Drilling Engineer must give the coring company all of the above information if they are to be able to deliver their best performance. He should also ensure that the hole is in good condition prior to coring and that all the necessary hardware required to handle the cores and fish the core barrels is on site. He should liaise with the Reservoir Engineers and Geologists to ensure that a mechanism is in place to deliver cut cores as speedily as possible to their analysis laboratory.

A decision must be made by the Drilling Engineer, the Drilling Manager and the Petroleum Engineers involved as to the length of core barrel that will be used. Traditionally, 30 ft core barrels were standard. These have grown over the years to the point where cores of several hundred feet are being routinely cut in forgiving homogeneous formations. Outstep well records will assist the selection of core barrel length to be selected.

3.3 Preparatory work prior to coring

3.3.1 Hole condition

The Drilling Supervisor should discuss any potential hole problems with the Coring Engineer prior to making up the core barrel on the rig. Any tight spots or dog-legs should be considered as potential trouble spots, especially if previous runs have been with Tricone bits on limber BHAs.

Note: Core barrels should be considered a stiff packed hole assembly and if there are tight spots or dog legs present, then a core barrel is much more likely to hold up than any other drilling assembly.

Water loss from mud should ideally be kept as low as reasonably possible (excessive water loss causes thick wall cake to build up which can cause coring problems). Special care should be taken to ensure that the hole is free from any junk prior to coring (run a junk basket on the last bit run before coring). Junk slots on coreheads tend to be smaller than on diamond or PDC drill bits.

3.3.2 Coring hardware

The following is a suggested list of hardware which should be available on site:

Equipment	Quantity required
Coreheads	3 minimum
Core barrels	2
Pup joints (for space out)	as required
Overshot (for fishing the core barrel)	1 + spare grapple for each size of barrel in use
Circulating valve	1
Coring balls	3
Spares and handling tools	1
Cross over (core barrel to drill collar)	1

It would be advisable to have the following available as backup, ready for immediate dispatch should this become necessary:

additional spares
additional core heads

specialised equipment for difficult formations, i.e., soft formation catchers, vented assembly etc.

3.3.3 Other material

The following should also be made available for the coring operation, if required:

wax and bath
foam
core boxes and trays
fibre tube
diamond saw
diamond plugging equipment.

3.3.4 Use of fibre glass inner barrels

Fibre glass coring is sometimes used when coring in soft, unconsolidated formations and although fibre glass inner barrels were originally introduced to cut 30 ft sections of soft formations, they are increasingly being used in longer sections. This means that systems must be utilised to break and layout each 30 ft section safely without breaking the fibre glass or dropping pieces of core during the operation. The essential points of handling fibre glass inner barrels are:

1. The section of tube being unscrewed must be stabilised by the racking arm or by someone on the stabbing board to prevent it swinging when the threads are undone.
2. Tension on the tugger-line should be kept to a minimum, otherwise the tube may jump.
3. The clamping device must be installed at the earliest opportunity.
4. Care should be taken in swinging the tube out of the V-door as a collision could fracture the fibre glass.

3.4 Coring practices

3.4.1 Make-up

1. Make-up should always follow the manufacturers guidelines.
2. Points to watch out for at this stage are the general appearance of the barrel, the ease with which joints make up. Without tearing the barrel apart, it is extremely difficult to make further inspections.
3. Before running in the hole, it is prudent to check that the safety joint has not been 'cracked' when making-up the first collar. It is not advisable to reverse in collars as this can lead to the loss of the outer barrel.

Tripping either in or out of the hole, should be carried out with some caution because of the possibility of swab or surge due to reduced clearance around a corehead compared with a Tricone bit, and Tight spots or dog-legs. This is especially important on the first core run. If problems are encountered, it should be remembered that diamond bits suffer disproportionate gauge wear while reaming and this kind of operation should be kept to a minimum.

3.4.2 Commencing coring

On reaching a position one or two singles off bottom or the end of the last stand, it is advisable to wash down to bottom any clear or cavings which may have accumulated. Check the position of the 'bottom' and check the length(s) of pup joint(s) needed to give the maximum possible run before having to make a connection.

Once the bottom has been tagged satisfactorily, the ball can be dropped when the pup joint(s) are added to the string. Circulation is then continued until the ball has seated, which is indicated by a sharp increase in circulating pressure. If fibre glass inner tubes are being used, a reduced circulation rate should be used to avoid shock damage to the fibre glass when the ball seats. When the ball has seated, slow-circulation rate measurements must be taken using both pumps as a precaution, in case a kick requires well killing procedures. When all these operations are completed, coring can commence. The flow rate should be set as required. Ranges for this are generally available in coring manuals but can be varied according to experience in a particular formation. The rotary speed is then set in the range 30 – 60 r.p.m., and the bit set gently on bottom until the initial jump in torque has smoothed out as the bit beds itself in. In soft formations this may occur virtually instantaneously, while more weight will be required for harder formations. The RPM and SPM should be altered to achieve the best possible performance. Each change should be small and executed smoothly. The changes should be on the following scale:

RPM	10 – 20 RPM
WOB	1 tonne

The first core is always somewhat of an experiment. Sight of the core can give vital clues as to any changes or modifications needed to improve performance, either in penetration rate, or in core recovery. It must always be remembered that core recovery is far more important than penetration rate, or indeed any other consideration apart from safety.

3.4.3 Circulation rates

The recommended ranges are given in coring manuals and should only be exceeded in exceptional circumstances. The exact value chosen for any particular core run is a considered judgment, taking into account the formation being cored, bit cooling and cleaning and hole cleaning needs. In general, the highest possible flow rate within the range indicated for a particular bit size and which does not adversely affect the core, is used. It is normal practice to maintain the circulation rate at a constant level during coring to enable clear monitoring of the corehead performance and condition.

3.4.4 Breaking off the core

Two occasions arise when a core must be broken, namely to make a connection and to end the core run. In both cases the procedures are the same. First, stop the rotary and then pick up the string until the core breaks or a pre-set maximum overpull is reached. This operation should be carried out very gently without unnecessary jerks.

Except in the case of extremely tough formations, such as silica cemented sandstone, the core will break well below the maximum listed for each core size in the manuals. For very tough formations, holding the core at the maximum recommended overpull and circulating for several minutes can usually induce a fracture and allow release. If this does not occur, rocking the table and inducing torque into the core can be tried before resorting to the final option of just pulling. In this case the overpull limit on the drillpipe and core barrel must be taken into consideration, but the most disastrous scenario is normally that the catcher can be pulled through the catcher shoe resulting in leaving the core on bottom. An attempt should be made to use the core barrel as an overshot to recover the core, as trying to drill up a core in this situation is extremely difficult.

Once the core has successfully been broken, the string should be picked up a minimum of 10 feet and run back slowly to within 1 foot of bottom to ensure that the core has actually been picked up and that there are no major pieces left on the bottom. If the core is then to be retrieved, it is normal to carry out a flow check, pump a slug and POOH. A careful watch should be kept for indications of swabbing, the trip should be carried out as gently as possible to avoid losing the core. Rotating the drillstring on the way out of the hole is not recommended.

If the break is to allow a connection to be made, the string should be picked up, the slips set gently and the connection made without rotating the drillstring. On restart, the core barrel should be run back to the bottom without rotation or circulation and weight on bit added to slightly over the previous coring weight to release the core catcher. In soft formations add 5 – 10 per cent over the coring weight and in hard formations add up 50 per cent over the coring weight, then pull back to required coring weight. The pumps should then be started and brought up to the correct circulating rate, starting the rotary table at the same time. Coring can then be continued.

If anything is going to go wrong, it will go wrong at a connection. The vast majority of premature trips during coring occur after an aborted connection, hence the importance of proper space-out measures before starting the run. If a core has jammed, it is unlikely that coring can be restarted and if restart procedures fail two or three times, it is time to consider pulling out with the core cut. Continuing to try to restart a core can cause the catcher and the catcher shoe to split, thereby losing the core.

Note: Always ensure that coring is stopped prior to overfilling the barrel as filling a core barrel to capacity will cause problems when extracting the core.

3.4.5 Extracting the core

Only coventional coring will be dealt with in this section as other techniques have specialised handling methods. The details of the operation are available in any coring manual. The salient point regarding the core is to ensure that it is placed the right-way-up in the core catching trays, with the minimum of breakage or spillage. Pieces of core should be raked up using a hammer or shovel. The line of vision between the Driller and the Coring Engineer should be uninterrupted at all times to ensure their actions are coordinated. The handlers must not put their hands or feet under the end of the inner barrel when extracting the core.

When the core is stuck in the barrel and cannot be shifted by hammering the tube, the core must be pumped out. The inner barrel is tied down on the cat-walk with the bottom end pointing away from any fragile equipment. The pump must be high pressure and low volume, thereby reducing the risk of rapid expulsion of the stuck core. It is attached to the top end of the tube and a rubber bung is used as a plunger. The pressure is built up slowly until the core begins to edge out of the barrel. Normally, the expansion is relatively slow and can be controlled by pressure release valves.

In areas where H_2S is present, the gas can be trapped along with the core in the inner barrel. Hand held sniffers can be used to test for the gas during the trip out when connections are broken, and when the ball is retrieved from its seat before the core is recovered. This equipment should also be used as the core is being dumped since the H_2S gas is heavier than air. When the gas is expected in harmful quantities, breathing apparatus should be available for all personnel required to work on the drill floor, as well as escape units for other personnel likely to be affected.

While the core barrel is at surface, the Coring Engineer must check all parts of the barrel and replace any which show signs of wear or damage. The items most often requiring changing are O-rings, thrust bearings, catchers, catcher shoes and valve ball seats. The corehead should also be checked for wear, damage and gauge retention. The decision can also be made to change the corehead to improve recovery or penetration rate.

3.5 Coring problems and indications

3.5.1 Jamming

Jamming is indicated by a sharp change in pump pressure and/or torque and/or rate of penetration. Any situation in which there are rapid changes in pump pressure and/or torque must be watched very closely and, if there is any doubt as to the source of the fluctuation, then the core should be pulled since the recovery of core is the prime aim of the whole operation. Sometimes a core can jam with very little or no indication and what looks like the characteristic torque and pressure fluctuations can be caused by formation changes. These kind of problems normally occur in soft or very fractured formations such as chalk. Experience in the area is the only effective way to predicting these causes and effects.

3.5.2 Sticky hole

Sticky hole problems are usually restricted to tripping. If severe, they must be treated by running a bit and conditioning the hole before attempting to core. Between core runs it may be necessary to condition the hole. If the core barrel is run in these circumstances, there is a likelihood of sticking and of burning the gauge of the corehead. The main problem caused by sticky formation while coring is torque on the collars or stabilisers which interrupts the smooth flow of power to the corehead which in turn can cause jamming.

3.5.3 Cavings

If a hole contains a heaving formation, it is essential to keep the pipe tally up-to-date, especially when pups and singles are being picked up and laid down, as hard fill can be confused for formation, if the pipe tally is in doubt. If cavings are present, they must be drilled out before the ball is dropped, and the hole must be clear before coring commences.

3.5.4 Washouts

In the case of suspected washouts in the core barrel or drillstring, the core barrel should be lifted off bottom and the off bottom pressure recorded at the start of coring, compared with the new pressure. If this is significantly lower, and the pump rate has remained constant or has increased, the drillstring and core barrel should be pulled out of the hole and checked.

Note: The difference between on and off bottom pressures for surface set diamond coreheads will be much greater than for PDC core heads.

3.5.5 Changes in penetration rate

Changes in penetration rate occur on crossing formation boundaries and are normally associated with torque and pump pressure shifts.

3.5.6 Junk

Junk in the hole, from whatever source, is extremely harmful to coreheads, especially to diamond coreheads, since the gap between formation and bit body is small and junk channels are normally very limited. Thus junk can be trapped below the bit. In soft formations it may still be possible to make progress, but bit damage will occur, normally scarring and incipient 'ringing'.

3.5.7 Lost circulation material

Coring can continue even with considerable quantities of LCM in the mud, but great care must be taken. LCM which is designed to mat, can cause blockages at the valve ball seat, between the inner and outer barrels and in the inner barrel. When running in the hole, the barrel must be circulated clear several times to avoid a build-up of material and circulation should be maintained for longer than normal before dropping the ball.

3.5.8 Corehead failure

Corehead failures occur for two basic reasons:

1. Hard material such as junk, pyrite or chert gets caught between the bit and the formation. This rolls around under the bit causing chips or scars which can then extend and become a 'ring'. The indications of this are initially low penetration rates and irregular torque followed by an increase in pump pressure and progress going to zero.

2. Soft material such as clay, LCM clogs the waterways on the bit. Indications of this are higher than normal on bottom pump pressure and low torque as the diamonds or Stratapax are starved of cooling fluid. The overheating leads to 'burning' and eventually to very slow penetration rates.

3.5.9 Bearing failure

The signs of bearing failure vary from no change to erratic torque as the catcher tries to grind the outer surface of the core while the inner barrel rotates. Most equipment failures result from stresses encountered when:

1. A core has jammed and excess weight has been transmitted to the inner barrel when attempting a restart.
2. High overpull is required to break a core.
3. Attempting to ream a long section of hole.
4. Sticky hole conditions or dog-legs resulting in stretching or bending while tripping.
5. Hole conditions result in high and erratic torque.
6. Cavings or junk get jammed inside the barrel.

4. MEASUREMENTS-WHILE-DRILLING (MWD)

MWD systems allow the following variables to be measured:

Wellbore surveying	azimuth inclination magnetic tool face gravity tool face
Drilling optimisation	weight-on-bit/torque-on-bit annulus and drillstring pressure temperature mud motor RPM bending moment bit pressure drop
Formation identification/ evaluation	gamma ray formation resistivity mud resistivity porosity/neutron log

(The purpose of these logs is explained on page 164.) The range of services provided by MWD are improving and the rate of change is accelerating. If the present rate of development continues, then MWD could effectively replace electric logging for many evaluation applications in the near future.

5. ELECTRIC LOGGING

In chapter 2, the properties of a reservoir were discussed. Briefly, the reservoir needs to be porous, permeable and sealed by a cap rock of sorts. Electric logging can indicate the porosity of formations, their thickness, the type of fluid occupying pore space and, by using wireline, run formations testing their permeability. Other specialised tools can indicate the bedding plane angle of the formations and the likely fluids in the pore spaces. This section discusses the selection and use of electric logging tools.

5.1 Determination of porosity

Three types of logs will indicate a formations porosity:

1. Sonic
2. Density
3. Neutron

5.1.1 Sonic logs

Somic logs measure the speed of sound across the formation being logged. Denser formations will give faster transit times, inversely indicating the degree of porosity. Compensated sonic tools use two sets of transmitters and receivers mounted on the tool and integrate their results. Sonic logs can also be used to provide formation velocity data which can support assumptions made in initial geophysical work. This helps identify formations and what is in their pore spaces.

5.1.2 Density logs

These are radioactive logs which transmit gamma rays into the formation. Using receivers on the tool, they measure the reaction of the electrons in the formation to the gamma ray bombardment. This measured reaction indicates the specific gravity of the formation which inversely indicates porosity.

5.1.3 Neutron logs (SNP/CNL)

These tools emit neutrons and measure how much these neutrons are slowed down by the nuclei of the formation material. Hydrogen is the most effective element in slowing down these nuclei and has the most effect on the log, regardless of what compound it is present in. Liquid hydrocarbons and water possess similar concentrations of hydrogen. Consequently, if the rock itself has minimal quanities of hydrogen in its construction and its porosity is totally filled with oil, condensate, or water, then porosity values can be accurately determined using neutron logs.

If, however, there is gas in the formation spaces, or if the rock itself has an appreciable amount of hydrogen in its make-up, then the porosity readings gained by using neutron logs will indicate a higher porosity than actually the case. By utilising this phenomenon and comparing neutron logs with other logs showing porosity, (sonic and density) an indication of what is present in the pore spaces can be derived.

5.2 Determination of lithology

Having identified potential reservoirs, we need to delineate their depth thickness and relationships with other strata. Two types of logs will indicate these lithological features:

1. spontaneous potential logs
2. gamma ray logs

5.2.1 Spontaneous potential logs (SP)

This tool records the natural potential of wellbore formations. Shale gives a very consistent level of spontaneous potential and is therefore used to provide a baseline against which other formations can be compared.

Formations containing freshwater in their pore space will show a higher SP level than those containing saltwater. These effects will be more noticeable in permeable formations, since the SP is measured by ionic flow and permeability allows ionic flow to occur more readily through the fluid in the pore space.

When hydrocarbons are present in the pore spaces, they will give a reduced SP level than that of the water-filled spaces. Using this phenomenon, oil/water contact levels can be estimated.

5.2.2 Gamma ray logs (GR)

The gamma ray log measures the natural radioactivity of the wellbore formations. It is typically used to identify shales which exhibit high natural radioactivity levels. As such, it is a very effective tool and can be used to locate formation tops and the amount of shale in formations with a good degree of accuracy.

5.3 Determination of formation fluid type

The formation fluid type can be indicated by using resistivity logs such as:

1. induction logs
2. laterolog

Resistivity tools work on the principle of emitting current from one part of the tool and measuring this current at a different point in the tool after the current has flowed through the surrounding formation

Electrical resistance by definition is that characteristic of a material that tries to prevent the easy flow of electrical current though the material. Since oil has a higher electrical resistance than water, identifying hydrocarbon bearing reservoirs should, theoretically, be easy. In practice, however, the resistivity tools have a finite range of measurement into the wellbore walls and, of course, the rock immediately behind the wellbore will be contaminated by the filtrate from the drilling fluid. The degree of invasion of mud filtrate will depend on formation porosity, permeability, mud hydrostatic overbalance and rheological properties (especially water loss and filter cake effectiveness).

When using freshwater muds, the filtrate will usually exhibit a higher resistivity than formation water. Conversely, when using saltwater-based muds, the filtrate will exhibit lower resistivity characteristics. (The presence, for example, of dissolved salts, such as sodium chloride, enchances the electrical conductivity of water and thus lowers the resistivity.)

These phenomena, coupled with not knowing in most cases what the formation water properties are exactly, nor the depth of invasion, make the interpretation of resistivity log results far from straightforward.

5.3.1 Induction log

These resistivity logs are used when drilling with either air, foam, freshwater-based muds or oil-based muds, since the induction system does not need to transmit current through the drilling fluid in use. Induction logging tools have transmitter coils which produce eddy currents by electromagnetic induction. These currents exhibit a magnetic field which induces measurable voltages in the receiver coils which are mounted elsewhere on the tool. The voltages measured by the receiver coils are affected by the resistivity of the wellbore formations and consequently provide information which can be interpreted to indicate reservoir fluid type. Most wells are, in fact, logged using a dual induction log. By measuring resistances at two depths into the formation an assessment of mud filtrate invasion can be made.

(This data is useful, not only as an evaluation tool, but also to predict potential well test performance by quantifying the amount of formation damage that might have been done by the invading filtrate. If the filtrate entered pore spaces in a pure form then flowing it out would not be too difficult. In practice, however, as the filtrate enters the pore spaces it takes drilled solids and mud chemicals with it. These have to be flowed back out into the wellbore during production tests if the reservoir is to achieve its full potential).

5.3.2 Laterolog

This logging tool has a current emitting electrode positioned inbetween two long guard electrodes. A constant current of electricity is supplied to the emitting electrode and an auxilliary current of similar polarity is supplied to the two long guard electrodes. These currents are maintained so that there is no potential difference between the current emitting electrode and the long guard electrodes. Because of this, the emitted current flows directly into the formation. This current is picked up by a receiving electrode mounted outside the emitting/guard electrode grouping. The potential difference between the emitting and receiving electrode gives an indication of the resistivity of the formation fluids.

This operating principle allows the laterlog to operate well in formation bands as thin as 50 cm. The induction log, in comparison, will only work effectively in formation thicknesses of 1.50 m or more. The laterolog tool works best in a saline mud environment since the mud filtrate from these muds has less effect on the natural resistivity of the formation fluids. As with the induction log, laterologs tend to be run as a dual unit which takes a shallow and a deep look at formation resistivities.

5.4 Other electric logging tools

Additional information can be gained by using any of the following tools:

1. dipmeter
2. formation tester
3. sidewall samples
4. casing collar locator
5. cement bond logs
6. caliper logs
7. free point indicator
8. temperature survey

5.4.1 Dipmeter

This tool measures the plane of the formation relative to the horizontal and the hole direction. The dipmeter works by measuring formation resistivity, usually through four arms. Resistivity is recorded simultaneously and the results gained indicate formation bedding planes. The tool also records hole angle, direction and caliper (see page 168).

5.4.2 Formation tester

Wireline formation testers need to have the following characteristics:

1. Selective setting so they can be placed alongside the formation to be tested and then pressed into the formation to provide an effective seal.
2. Sample chambers in which to take formation fluid samples.
3. Pressure measurement by which reservoir pressures/build up rates can be recorded.

The most commonly used formation test tool is the repeat formation test tool (RFT). This can be set and unset gaining information from several zones. The RFT is an open hole tool, but it is possible to carry out formation tests through casing by using a formation interval tester tool (FIT). The FIT is equipped with perforating devices capable of shooting a hole through casing to take pressure readings and samples from the formation behind the casing and cement.

5.4.3 Sidewall sampler (SWS)

The sidewall sampler tool is equipped with small steel cups mounted on a frame from which they are propelled by an explosive charge. The cups are forced into the formation at the shooting point and then pulled out by the tool as it is pulled clear by means of a steel wire which attaches the sample cup to the tool. The sidewall sample tool can, theoretically, be of any length, however, a typical tool will have 30 cups, all of which can be selectively shot into the formation.

Clearly, the value of the information which can be gained from the sample thus obtained will depend on the depth to which the cup penetrated, the amount of filter cake present and the loss of formation material, if any, during the trip out of the hole. With a little luck though, it is possible to identify formations and to quantify their porosity and permeability from the SWS results. SWS results can provide base reference material for other logs and, done selectively and successfully, are quicker and cheaper than conventional coring methods for achieving the same results.

5.4.4 Casing collar locator (CCL)

The CCL as its name implies, locates casing collars or any change in steel thickness. This tool is used when selecting shooting depths for perforation or for setting packers. It is used in conjunction with the gamma ray tool which can operate through casing to identify formation tops. It is common practice to correlate the CCL log with the casing or tubing running list to confirm that everything is where we thought we put it. The CCL works on the same principle as the gamma ray tool.

5.4.5 Cement bond logs (CBL/VDL)

Cement bond logs are sonic tools which emit sounds and pick up reflections. They are used inside casing as a means of locating the top and the quality of cement behind the casing. Uncemented casing resembles a tubular bell and transmits sound easily. Well-cemented casing will dampen sound vibrations and produce different results which can be interpreted together with well data. The value of CBL/VDL is debatable. The author has seen apparently uncemented casing behind which communication was impossible up to formation leak-off pressures. Conversely, he has also seen beautiful CBL/VDL logs on wells with live annuli. Bearing in mind the above reservations, the tool should be used as a guide only and, when well safety is in question, an alternative means of establishing whether or not a seal exists should be utilised.

5.4.6 Caliper logs

Caliper tools use arms which extend out from the body of the tool to determine the hole diameter. They range from two-arm to four-arm calipers and clearly, the more arms that you have on the tool, the closer to the actual hole configuration the results will be. The commonly used tool today is the borehole geometry tool which utilises four-arms. Modern logs such as this will not only plot hole diameters but will totalise hole volume against depth and even print out cement volume requirements, given a quoted casing size to be run in the logged hole. In addition to using this log for hole volume calculations, it is also used to select casing seats for open hole packers and to indicate degree of washout or reaction to drilling fluids in the hole.

5.4.7 Free point indicator (FPI)

This is a tool which is usually run inside drillpipe to detect the point at which it is stuck. It has a set of pads mounted at each end of the tool and a means of measuring stretch and torque between these pads once they are set against the inside wall of the drill pipe. Stretch and torque applied to the drillstring at surface will register on the tool unless the tool is positioned below the stuck point. Thus, by moving the tool progressively down this hole and applying stretch and torque, the stuck point can be determined. In most stuck pipe situations, the Drilling Supervisor can predict with reasonable accuracy the point at which the string is stuck. In these circumstances, the FPI need not necessarily be run.

5.4.8 Temperature survey

This log measures the temperature of the borehole fluid at a given depth. If the log has been run earlier in the well's history, most tools are capable of comparing present temperatures with those previously recorded and to print out this differential temperature on a continuous basis. The tool is used to pick up tops of cement as heat is generated during the hardening process. It can also be used to determine geothermal gradients and to indicate tubing and casing leak points.

5.5 Combination logs

As most wells require several logs to be run to fully evaluate their reserves, much thought is given to combining logs into single run tools. In chapter 3, (F. 27), we state that radioactive logs should never be the first logs run in a sequence. Apart from that restriction, we can combine the logs up to the limit of the tools' capabilities. A typical combination of logs would be as follows:

Run 1	sonic, laterolog, gamma ray, SP, caliper
Run 2	density, neutron, gamma ray, caliper
Run 3	dipmeter
Run 4 etc.	depending on wells needs

By running logs in this fashion, time can be saved and results still achieved in an acceptable format.

6. WELL TESTING

6.1 Objectives

Well testing is generally carried out to provide the following information

Identification of formation fluids
Determination of reservoir pressure and temperatures
Confirmation of electric logging interpretation
Estimation of formation damage (skin effect)

6.2 Types of well tests

Well tests fall into two main types:

1. Drill stem test (DST)
2. Production test (PT)

6.2.1 Drill stem test (DST)

Originally, the objective of a drill stem test was to take a representative fluid sample which, in the very early days of the oil industry, was obtained by either bailing or swabbing. Following the introduction of a packer and a downhole valve, the next step was the possibility of recording accurate bottomhole reservoir pressures and temperatures. With the development of downhole tools, which operated on annulus pressure and a downhole shut-in valve with surface read-out, drill stem testing has become a generally accepted safe method to meet the original objectives: to obtain a fluid sample and reservoir data.

In a drill stem test, information on the following parameters can be obtained:

1. Reservoir fluid type and properties
2. Production rates and potential production problems
3. Initial reservoir pressure
4. Effective permeabilities
5. Type of flow system

A drill stem test is the temporary completion of a well to gather reservoir data. The three fundamental components are a retrievable mechanically set packer, a downhole valve assembly (DST tools) and drill pipe/tubing to surface.

A drill stem test is made by running a bottom assembly, consisting of a mechanically set packer and a tester valve, on the drillpipe or tubing. The tester valve is closed while the drillstring is run, thus pressure inside the drillpipe tubing is low compared to the hydrostatic mud or brine column pressure. To provide some back pressure, the string is usually partially filled with water (water cushion). Once on bottom, the packer is set to isolate the zone to be tested from the mud column and the tester valve is opened by pressuring the annulus to allow formation fluids to enter the test string.

After a suitable testing period, the valve is closed by releasing annulus pressure and a pressure build-up occurs below the valve as formation fluids repressurise the area around the wellbore. After an adequate build-up time, additional flowing periods can be made by opening the tester valve again. Before pulling out, a circulating valve is opened by pressurising the string 500 p.s.i. above the tester valve operating pressure. The string content can then be reverse circulated out and the well can be killed.

The tool string is usually equipped with very accurate pressure and temperature recording devices and a complete flowing build-up record can be obtained which permits determination of reservoir pressure, permeability and sometimes formation damage.

Open-hole testing is the cheapest possible test since it does not require the use of casing, cementing and perforating. An open-hole packer is set in the open hole at a point where the formation is competent enough to allow it to seal. Drillpipe is run in the hole, either empty, or with a water cushion and, by manipulating the string valves, the formation under test is allowed to flow into the drill pipe. Due to the strong manipulation required, this type of test is only suitable for land rigs and fixed offshore structures where rig motion is not a problem.

Barefoot testing is similar to the open-hole test with the exception that the packer used is set is casing and not open hole. A stinger is hung off below this packer to reach the zone of interest.

In **Cased-hole testing**, the test duration is hardly affected by chances of sticking as in open-hole or barefoot testing. Also packer seal capability is enhanced and thus, the test design can be more flexible. Both string manipulation and annulus pressure activation can be applied to operate the downhole tools, pending rig type and casing conditions. The disadvantage, of course, is the additional cost of a string of casing and the rig time involved to run and cement the casing in place.

6.2.2 Production testing (PT)

A production test is flowing a well through tubing and surface equipment in a temporarily completed well. Its primary purpose is to collect information to predict well and reserve oil performance under future operational conditions. The well is produced according to a schedule of predetermined rates, during which bottom hole and tubing head pressures are recorded and produced fluid quantities are measured. In the course of the test, wire line formation tests (WLFT) can also be taken.

Depending on the test objectives, the duration of the test may range from a few days to several weeks/months. The PT method is often considered to be very time-consuming and expensive, but those concerned with the planning and execution of the test must always judge the duration and cost of testing in relation to the potential rewards and to the overall venture costs. Often, complete production stations and, sometimes, very costly offshore platforms, are designed on the basis of one or two production tests. In either context, the cost of a few days testing might be quite modest. It may be observed that the time and money spent production testing is a good investment ensuring the profitability of a field development.

6.3 Test string designs

6.3.1 Test string requirements

The modern test string is called on to fulfil a number of other requirements during the well test. Sometimes, these facilities already exist in a particular tool or another tool may have to be added to the string. Some of these special requirements are listed below:

1. Sampling
Bottomhole sampling may be taken during the test for analysis. The sample can be caught in the downhole valve and recovered with the rest of the string.

2. Safety valves
A downhole safety valve, independent of the main valve is often required to provide additional well control, especially if the main downhole valve fails. These valves should be operated by a different means from that of the main valve.

3. Spotting tools
Included above the main valve, this is a specialised tool which may be required to spot (or selectively place) fluids such as acid or injection water. This is done to avoid bullheading into the formation and thus a reclosable feature is required. The spotting tool is usually combined with a reversing tool.

4. Expansion/contraction
In most tests (especially with pressure controlled tools and offshore floaters) some string expansion and contraction has to be compensated for. Slip joints or locator seal asse ıblies perform this function.

5. Data monitoring
Onsite data validating requires the capability of real-time monitoring of bottomhole pressures and specialised tools are needs to provide this capability.

6. Full bore
In order to allow direct access to the information with wireline and coiled tubing and allow high flow rates and stimulation treatments, an unrestricted flow path (ID) is required. The individual tools can be used in other strings, for example production tests, if a particular need exists.

7. Tubing conveyed perforating (TCP)
When TCP is run, the type of firing system used will be influenced by the downhole test string and vice versa. The string may have to be designed around the TCP.

6.3.2 Drill stem test string design concepts

A DST string must conform to certain basic principles regarding the type of tools required, their function and where they need to be placed. A DST string in the most basic form, consists of:

- a conduit to flow the well to surface
- a packer to isolate the annulus from the formation below the packer
- a downhole control valve for well control, downhole shut-in and pressure testing
- a circulating valve to reverse out conduit content

The basic DST string applies to all types of offshore installations, floating rigs (both semi-submersibles and drill ships) jack-ups and platforms. Extra equipment for floating rigs, is required and is covered on page 173.

Basically, three pressures need to be considered in test string design:

1. Formation pressure (Po)
2. Hydrostatic pressure (Hh)
3. Cushion pressure (Pc)

Hydrostatic head must always be available to control Po. The basic test string consists of the tubing or drillpipe as a conduit for flowing the well to surface, while still maintaining the hydrostatic head between the test string and production casing. The two pressures Po and Hh must be isolated from each other in order to allow the well to flow and this is achieved by having a packer positioned near the bottom of the test string. For a DST, the packer used would normally be retrievable.

The packer and the string are the two most basic elements of a test and very simple operations can be conducted only with their use. If this is all that is run however, the string will fill with mud and it is this hydrostatic head that must be relieved to allow the well to flow. This can be achieved by swabbing, or by pumping a lighter fluid into the string to reduce Hh below Po, just before the packer is set. These methods are used, but potential well control is reduced and they are not always successful. Therefore, to allow running of the test string with a lighter cushion fluid, a downhole tester valve is installed in the test string.

6.3.3 Production test string design concepts

A production test string must provide the same capabilities as that for a DST. However, the concept of production testing differs in that the string is a permanent type of completion. The major difference between a DST string and a PT string is that a permanent packer will be used in a PT. A seal bore extension is normally used and a locator seal assembly is stabbed into the packer with a seal extension to effect a seal.

6.4 Testing on floating drilling drilling rigs

The basic differences that exist between fixed drilling rigs and floating drilling units are:

1. Motion or heave of the drilling rig
2. Production casing protection stops at the seabed and remotely controlled
3. BOP equipment is located at the seabed and remotely controlled

To solve these problems, the subsurface test tree (SSTT) was developed. The SSTT is landed in the wellhead and thus is a fixed point below which no rig movement is transmitted. It also provides both well control and unlatching capability in the event of severe weather or dynamic positioning problems. Following emergency closure of the SSTT and disconnection of the hydraulic operator, the blind rams can be closed above the master valve during temporary abandonment.

The SSTT assembly consists of a fluted hanger, slick joint, valve section and latch section. The fluted hanger is designed to land on the wear bushing in the wellhead. Above it is the slick joint spaced out so the lower pipe rams can close around it to seal off the annulus. The valve section contains two fail-safe valves, either a ball and flapper valve, or two ball valves. On top of the SSTT is the hydraulic latch section. This contains both the operating mandrels to open the valves and also the latching mechanism that releases this part of the tree from the valve section, in the event that disconnection is necessary.

Control of the SSTT is achieved by three control lines in an umbilical hose connecting the hydraulic latch section with the surface control manifold. These three hoses provide hydraulic pressure to:

1. Hold the master valve open – control line

2. Assist the valve in closing – balance line
3. Unlatch the hydraulic operator from the valve and test string

A fourth hose is usually contained in the umbilical for use as a chemical injection line, such as methanol injection, for hydrate inhibition when testing gas wells. The complete hose is contained on a reel unit which automatically controls pay-out or take-up via a self-contained air driver motor. The hose bundle is connected from the reel to the surface hydraulic control manifold. This is a complete, unitised assembly, usually containing a hydraulic reservoir, pressure-controlled regulator, gauges, gas powered pump and relief valve. The unit is designed to maintain a predetermined pressure on the normally closed SSTT safety valves, to hold them open. It also supplies control pressure to activate the release mechanism to unlatch the tree. A separate chemical injection pump would be provided if chemical injection at the SSTT was required.

6.4.1 Deep water testing operations

The use of anchored semisubmersibles and more particularly, dynamically positioned semisubmersibles and drill ships has led to testing operations being performed in deep water. The testing of wells in deep water requires very long hydraulic lines to control subsea test trees. The response time required for valve closure and disconnect, previously adequate in relatively shallow water depths, would have proved far too slow in deeper water. This is because the time required to bleed down these long lines of hydraulic pressure is considerable and would delay the closing and unlatching of the SSTT in an emergency.

The result has been the development of an accessory package, mounted on the SSTT to operate in water depths greater than 1000 ft. The system, when properly installed, is designed to act as a 'quick exhaust valve' by discharging into the riser pipe and not returning to surface. This allows the SSTT safety valves to close in seconds as compared with the time to bleed off thousands of feet of control line in the conventional system. It is also designed to provide immediate high pressure at the SSTT location, allowing the latch to disconnect in seconds.

This system is controlled from the drilling vessel and the package consists of a deep water actuator, surface hydraulic control manifold complete with high pressure accumulators, a reel unit and control lines. The deep water actuator is mounted on the subsea test tree in the BOP at the ocean floor. It contains a bank of accumulaters, two servo valves and piping. One servo valve is designed to bleed off pressure from the SSTT.

Safety valves and the second valve are designed to release high pressure from the accumulator to unlatch the tree. The accumulators are continually charged from surface through a high pressure line.

6.5 Surface equipment

The surface equipment used in drill stem testing is common to all offshore rigs and consists of control, processing, measurement, sampling and safety equipment. Since drill stem tests vary in complexity, duration and purpose, a variety of surface testing equipment can be considered. The equipment selected for each test programme must be able to achieve the test objective.

6.5.1 Surface flow head (surface test tree)

The surface flow head is the primary closing and safety assembly located on the surface. The surface flow head is attached to the top of the test string, sufficiently high above rotary table to accommodate rig heave. It should have a pressure rating equal to or greater than the maximum expected surface pressure, and have sufficient tensile strength to pick up the weight of the test string to unseat the packer. The assembly consists of various components and the components selected for each test depend on the downhole tools, drilling rig and the test objective. The following components may appear on a surface flow head:

1. A lift nipple which provides a means of latching the elevators onto the test string above the surface control head.
2. An adaptor sub which can be used in place of the lift nipple so that a joint of drillpipe or 'pup joint' can be installed on top of the control head.
3. A bar drop sub can be used to hold and release the bar that shears the pin if an impact reversing sub is run.
4. A flo-tee contains side ports that allow flow to exit from the drillpipe to the flow line or kill line. In a $3^{1}/_{6}$ inch 10 000 p.s.i. WP flow head, the three valves (swab, flow and kill wing) are in a solid block system.
5. A swivel is included in the surface test string when downhole tools are operated by the rotation of the test string. The swivel permits the test string to be rotated. The control umbilical for the SSTT is wrapped around the test string to accommodate rotation.
6. The master valve is the in-line shut-off valve located below the Flo-Tee. It should be full bore valve and should allow rapid closing.
7. A swab valve is installed to provide vertical access to the test string.
8. An access valve can be run above the master valve(s). The access valve provides an injection port to permit injection of alcohol or chemicals into the flow stream. This is particularly important with gas wells and with wells in cold climates. It also provides a pressure gauge port on the test string.
9. A kill wing valve is a manual valve which is hooked up to the cement unit with an in-line check valve. The kill wing valve should remain open during testing operations to allow immediate (bullhead) killing of the well, if required.
10. A flow wing valve is the hydraulically actuated fail-safe valve that permits immediate shutting-off of the flow at surface. It can be operated from the rig floor, or from locations remote from the rig floor.
11. Wireline lubricators, when used, should be of sufficient length to handle the longest tool to be run. The lubricator should be pressure rated to withstand the maximum expected surface pressure. The lubricator is rigged up on top of the flowhead (the lift nipple removed).

6.5.2 Instrument manifold (data header)

An instrument manifold or data header is a length of pipe ported to receive various probes and measuring instruments. The instrument manifold should be rated to the maximum expected surface pressure. A dead weight tester (manual or electronic) measuring wellhead flowing pressure and a thermometer for wellhead flowing temperature can be installed. Wellhead samples can be obtained at the manifold and a port for chemical injection is provided. A sand detection probe, if required, would be installed in the data header. The instrument manifold should be located as close as possible to the surface control head and upstream of the choke manifold.

6.5.3 Choke manifold

A choke is used for regulating flow and pressure during the test. The choke manifold should contain full-opening isolating valves and be equipped with an adjustable and a positive choke. Bypass piping should be incorporated in the choke manifold assembly. Pressure and pilot taps should be available on the manifold. The manifold pressure rating should be equal to or greater than the maximum expected surface pressure.

6.5.4 Surface flowline

A 3 inch (nominal), usually 10 000 p.s.i. working pressure, steel line connects the Christmas tree to the fixed flowline. This size steel line is available for low temperature and sour service with Weco 1502 unions fitted. On floating rigs, care should be taken to ensure that the barge heave movements are accommodated by the steel line. A flexible hose is often preferred to the flexible steel line on floating vessels, because of vessel movement. Hoses with 3 inch ID, standard lengths of 30 ft or 50 ft and working pressure of 10 000 p.s.i. are available.

6.6 Processing equipment

The equipment requirements and operating parameters for processing equipment such as heaters, separators, metering equipment and pressurised tanks and piping are discussed in this section.

6.6.1 Flowline heater

A heater is used to increase the temperature of well effluents in order to prevent hydrate formation, paraffin depositions, or to assist in increasing the fluidity of produced fluids for easier separation. A choke is installed, as an integral part of the heater, to provide a heated pressure drop position. The heater should be rated to the maximum expected wellhead pressure. Standard pressure rating for heaters are 5000, 10 000 and 15 000 p.s.i. working pressure.

6.6.2 Separator and metering equipment

A separator is used to separate oil, water and gas from the stream flow so that they can be individually measured and sampled. The separator should be sized according to the flow rates expected. All separators must be equipped with two relief valves and should be properly vented.

Incorporated with the separator is the metering equipment for measuring the oil, water and gas. A range of orifice plates should be available on each test for measuring gas. Positive displacement or rotary meters are used for metering oil and water. A shrinkage meter or atmospheric chamber may be included with the metering equipment. A manifold is installed upstream of the separator, or constructed as for the separator, to control the direction of flow. This permits the separator to be bypassed during clean-up or reversing, and permits the flow of oil from the separator to the gauge tank, transfer pump or burner.

6.6.3 Pressurised tank

The pressurised tank may be used for initial well clean-up and/or to check the metering of liquids. Normally, the tanks are 100 bbl capacity and 50 p.s.i. working pressure. The tank also serves as a holding chamber and sample point. The tank should have sight glasses and a gas meter. A transfer pump is required to dispose of the liquids.

6.6.4 Piping

Piping lengths between the surface control head and floor choke manifold should always be kept to a minimum and have a working pressure greater than the maximum expected surface pressure. The piping can be either steel pipe with swivel connections, or flexible pipe, which allows the test string to be moved up and down during the test. Normally, flexible pipe is used for the flowline. The piping should be sized so that the internal diameter is larger than the largest anticipated surface choke size. Downstream of the choke manifold, the piping to the separator and burners should have at least the same working pressure rating as the separator. If a heater is used, the pressure rating of the upstream piping should be equal to or greater than maximum expected wellhead pressure. On many marine drilling units, much of the pipe is permanently installed.

6.7 Disposal/removal equipment

Disposal equipment requirements will vary depending on the location of the rig and local environmental regulations. Normally, the oil and gas is flared off through burners. Some offshore locations do not permit oil burning due to environmental constraints and so the oil is flowed to a tank or barge to be transported ashore. In production tests it is becoming common for the hydrocarbons produced to be pumped to a storage tanker for transport ashore and so as to recoup some of the initial financial investment.

6.7.1 Atmospheric tank

The atmospheric tank is needed for holding fluids and allows approximate volume determination to check meters. Size range from 50 bbls to several hundred bbls. Venting and diverting from the tank are considerations for planning. A transfer pump is needed to remove fluids for disposal.

6.7.2 Burner and boom

The introduction of the crude oil burner in 1969 was a major contribution to the development of extensive offshore well testing. The burner allows relatively pollution-free disposal of produced hydrocarbons on site. During well testing operations, the risk of storing non-stabilised effluent is removed and uninterrupted oil flow is permitted. The burner is made up of one or several combustion heads. The presence of solid particles in the effluent does not affect burning operations and the head is designed to burn crude oil, oil-based mud and foam. Each head includes:

1. An atomiser obtains efficient combustion of the effluent by reducing it to very fine droplets which avoid heavy black smoke and fall-out of unburnt particles. The atomisation is achieved by using either the mechanical energy resulting from the effluent pressure, or by supplying additional pneumatic energy from either gas or air available on the rig. The final atomisation is the result of these two processes, the former predominating at high flow rates, the latter at low rates. To avoid plugging, the atomiser is designed to allow the passage of solid particles of up to ¼ inch diameter. The atomiser is easily replaced (quick-union).
2. A cylindrical hearth stabilises the flame and facilitates the circulation of combustion air.
3. A pilot flare system is installed and comprises a gas pilot light fed with propane, butane, or gas from the separator and a remote-controlled electrical igniter unit with spark plug. As an option, a diesel oil pilot flare, fed from an air driven pumping unit, can be supplied.
4. A water rig (nozzles) injects water inside the flame. This controls the formation of heavy smoke and decreases the radiated heat.

Other components are the supporting fame, swivel joint and rotation system. The supporting frame supports the atomiser, hearth, piping, swivel and pilot light system. For safety the oil, air, gas and water are supplied through check valves to a vertical swivel joint distributor which enables the burner to be oriented to suit the wind direction. The supporting frame is mounted on a rotation system, again for burner orientation. The boom provides support for the burner and the lines, allows orientation, and limits heat radiation on the rig to an acceptable level. The boom is normally 18 m long and made up of several lightweight corrosion resistant hollow sections. A rotating base plate allows horizontal rotation. Piping is installed for oil, water, gas/diesel oil pilot flare and an electric line for the igniter. Additionally a 4 inch gas flare line with flame arrestor is usually provided. Gas is normally vented or flared through open piping.

6.7.3 Air supply

There have been incidences where gross contamination of an offshore installation has occurred using testing. This has led to an explosion and fatalities. As a result, the use of dedicated air compressor(s) to supply combustion air is recommended. It is for this reason that the Drilling Operations Policy Document (see chapter 3 section H.9) insists on the use of separate air compressors to supply air to test equipment.

Chapter 9

RIG SELECTION

Rig selection is the process whereby we define our hardware needs to fulfil the well objectives. In safe cost-effective drilling operations, it is vital that rig selection be given the highest priority and not left to chance. Section 1. discusses the general principles involved, whilst sections 2. and 3. look at land and offshore rigs, respectively.

1. GENERAL PRINCIPLES

The rig requirement can be defined within the following groups:

1. Type of rig
2. Lifting capacity
3. Substructure/pipe setback load capacity
4. BOP size, type, number and rating
5. Pump output and rating
6. Tank capacity
7. Solids treatment equipment
8. Storage capacity
9. Accommodation

Once the well plan is known, the requirements for the above groups can be established and a rig specification drawn up.

2. SPECIFYING A LAND RIG

Using the groups outlined above as general principles, land rig selection will be done as follows:

2.1 Type of rig

For land wells the following choices are available:

trailer rigs
package rigs
microdrill-type slim hole rigs

For wildcats, in most cases, the trailer rig will prove to be the most cost-effective since its mobilisation cost will be lower than the other types of rigs. The main disadvantage of this type of rig is its lifting capacity which, for some wells, will be insufficient. Microdrill-type slim hole rigs initially seemed to be 'the answer to a maiden's prayer'. However, they are not now used very much at all, mainly due to their cost (which is similar to a full-size rig's day rate) and to their restricted lifting capacity in some cases. There are cases, however, where slim hole rigs will be the best rig for the job. This will usually be in extremely environmentally sensitive areas, where no horizon pollution can be tolerated and, where land costs are so prohibitively high that the small footprint of this type of rig creates large savings for the operator.

The package-type of rig will be used on deeper wells and/or for development drilling on clusters where the initial mobilisation cost is being amortised over several wells and rapid movement between cluster cellars can create savings for the operator. In acoustically sensitive areas, anechoic rigs will have to be used.

To summarise:

Trailer rigs:	First choice for wildcat wells. Limited lifting capacity.
Package rigs:	Higher mobilisation cost. Suitable for deep wells and cluster drilling.
Slim hole rigs:	Use only in environmentally sensitive areas where limited lifting capacity is not a problem.

2.2 Lifting capacity

The rig must be able to handle the heaviest drillstring or casing string with a margin of safety. It is not just the hanging weight of the string that must be considered but also the drags and overpulls necessary to free the pipe if it did get stuck.

Normally, the heaviest string handled by a rig is the production casing string. As a rule of thumb, the lifting capacity requirement of the rig should be considered as the heaviest string to be run plus a 25 per cent margin of safety. The string weight should be calculated 'in air' and no buoyancy factor included in these figures. This is because if the string gets stuck, the hydrostatic upthrust or buoyance of the string is lost and to pull it free will require pulling more than its dry weight to free it.

It is acceptable to have a rig which requires stringing a couple of extra lines into the blockline system to meet the above criteria. If this is proposed, then the additional time taken to string extra lines should be costed, as it is possible that a heavier rig with a higher day rate could work out cheaper overall.

2.3 Substructure/pipe setback load capability

This must be sufficient to take the full drillstring stood back in the derrick whilst the heaviest string of the casing is being run. Again, in most cases, this will be the production string. It is not acceptable to have to lay out drillpipe to allow casing to be run as we never know when the casing might need to be pulled and the hole reconditioned with a bit and BHA.

2.4 BOP size, type, number and rating

In chapter 3, we defined minimum BOP requirements for each well type. With this basic policy and knowing the well plan and expected pressure regime, the BOP requirement can be specified. Since land rigs typically use smaller BOPs than those used offshore, extra care must be taken to ensure that casing and tubing hangers can pass through with reasonable clearance.

2.5 Pump output and rating

For a given hole size, a given calculation rate is required. The normally used minimum annular velocity is circa 30 m/min. This corresponds to 500 gall/min or 1900 l/min) pump output in 12¼ inch hole using 5 inch drillpipe. The pressure loss in a given circulating system can be calculated using a hydraulic slide rule or by computation. By knowing the hole sizes and casing setting depths, the pump output capacity and pressure rating can be defined. The rig pumps, therefore, should meet these requirements as a minimum acceptable level.

The bit nozzle pressure loss is a major part of the overall pressure loss in the circulating system and the use of PDC bits, which can work efficiently with lower bit hydraulic horse power, have reduced the HHP requirements for given hole sections. The mud weights to be used will also influence the HHP requirements and these must be considered. As a guide, the following pump requirements can be used:

1. Hole: 12¼″ to 500 m 8½″ to 1000 m – 500 gpm and 1500 psi
2. Hole: 17½″ to 500 m 12¼″ to 2000 m – 800 gpm and 2000 psi
3. Hole: 17½″ to 1500 m 12¼″ to 3000 m – 1000 gpm and 3000 psi

2.6 Tank capacity

The surface mud tank capacity must be sufficient to allow continuous treatment of the mud prior to pumping it downhole. Again as a guide, the following minimum surface tank volume requirements can be used:

1. Hole: 12¼″ to 500 m 8½″ to 1000 m – 60 m^3/380 bbl
2. Hole: 17½″ to 500 m 12¼″ to 2000 m – 90 m^3/570 bbl
3. Hole: 17½″ to 1500 m 12¼″ to 3000 m – 120m^3/760 bbl

2.7 Solids treatment equipment

Before considering this requirement, we must determine what type of mud systems are to be used on the wells to be drilled. Furthermore, the ease with which liquid waste can be disposed must be assessed. Simple water-based mud systems can be kept in shape by extensive dumping and diluting. This process takes some load off the solids treatment equipment, however, it does fill up the waste pit rather quickly. If oil-based muds are in use, then dumping and diluting is not an effective continuous treatment. Consequently, a full set of solids treatment equipment will be required.

Most land wells can be drilled efficiently with the following equipment when using water-based mud:

2 double-decker shale shakers	
1 de-sander	1700 l/min/440 gpm
1 de-silter	1700 l/min/440 gpm

If oil-based mud is to be used, then in addition to the above mud cleaners, a centrifuge should be installed.

2.8 Storage capacity

Under storage capacity, we must consider all the consumables involved in drilling a well. To operate efficiently, the rig must be capable of keeping the following materials close at hand:

water
sack chemicals
bulk chemicals
cement
tubular goods

Once the basic well design has been completed, a list of the above requirements can be made. No two sites are the same since they have different logistic needs. If mains water is available and the supply can be relied upon, then the water storage requirements are minimal. If the well is located in an 'oilfield area', then chemicals can be procured as needed and not just kept on site. Similarly, casing can be brought to the site only a few days prior to running.

2.9 Accommodation and offices

Company policy will, in most cases, dictate the level of supervision used on the drill site. Most land wells will have an on-site Senior Drilling Supervisor and Wellsite Geologist. Some will also have a Drilling Engineer and possibly a Night-shift Drilling Supervisor. Each of these must have a place to work, a desk, filing cabinets and shelves to keep files etc. They also need shower facilities, sleeping quarters and a small kitchen where food can be prepared.

Different companies have different policies regarding whether the Drilling Supervisor can leave the site during drilling operations. It is the author's contention that the Supervisor should never leave the site during these operations. If this policy is adopted then the need for good kitchen facilities is increased.

The accommodation and office requirement for rig selection purposes will be that which meets the particular well's and operator's needs. This can be easily defined once the basic programme has been prepared.

3. SPECIFYING AN OFFSHORE RIG

3.1 Type of rig

For most offshore wells, the type of rig required will fall into one of three groups:

1. Jack-ups, where the water depth is less than 100 m.
2. Semisubmersibles in hostile waters too rough for drill ships and in depths over 100 m.
3. Drill ships in calmer waters with depths over 100 m.

There are certain situations where any of the three types could be used to meet a wells' needs and in these cases each well approach should be costed out to find a clear economic leader. If there is no economic leader, then oilfield common sense or intuition will probably show you the way to go.

3.1.1 Jack-ups

Jack-ups are used for most of the offshore exploration drilling worldwide. They fall into two main categories determined by the equipment used on the bottom of the supporting legs. By far the most common type of jack-up is the 'spud can' type. This rig has spud cans or tanks mounted on the base of the legs. These are usually round and are equipped with a jetting system to allow them to be jetted into the seabed. Additional jetting systems are installed on the top of the spud cans to allow the can to be pulled out of the formation, should penetration into the seabed be too great. The second type of jack-up is 'mat' supported. This type of rig has a large single mat connected to the base of each leg. This is lowered with the legs and, since it has a larger area than individual spud cans, it applies less pressure on the seabed for a given jack-up weight. Consequently, it is used mainly in areas where the seabed is very soft and spud cans would penetrate too far into the seabed if they were used, or where seabed pipeline congestion precludes the use of spud can type jack-ups. Only when operating in these types of areas should mat supported jack-ups be considered. In all other applications use spud can type jack-ups.

Most modern jack-ups are of the cantilever design. This means that the derrick and substructure can be skidded inboard on the rig for rig tows and skidded out over a cantilever during drilling operations. The safe working load of the derrick and substructure is affected by the amount of extension along the cantilever that is being used. Consequently, when lifting, substructure and setback capacity requirements are being considered for cantilever jack-ups, the position that the derrick will be on the cantilever must be known or fixed.

3.1.2 Semisubmersibles

Semisubmersibles are used mostly in the range of water depths of 100 – 300 m. Beyond this depth, anchor lines pose a problem and although dynamically positioned semisubmersibles do exist, they are not used very much, due mainly to high operating costs and a lack of total confidence in their DP systems when used in hostile environments.

In areas where water depths exceed 100 m and heavy weather can be expected then semisubmersibles must be used. The choice will, in most cases, come down to using either second generation or third generation semis.

Second generation semis appeared in the mid-seventies and incorporated a lot of the lessons learned by the first generation of semis and submersible rigs. A major influence in their design was the increasing exploration activity in the North Sea. To make a rig pay in the North Sea meant operating it for 12 months of the year. This in turn meant that the rig had to be able to withstand 100 knot winds and 20 m seas on a regular basis. Since the North Sea locations were remote and supplying rigs was problematic due to the prevailing seas, deck loads were also increased to allow more equipment to be carried on board. Typical second generation semis are rig designs such as the Aker H3 or Sedco 700. These rigs typically have a variable deck load of 1500–2000 tons and can be moored in up to 300 m of water.

Some second generation semis have been 'upgraded' over the years usually by the addition of extra columns from the submerged pontoons up to the main deck. This modification has the effect of increasing the variable deck load which in turn allows the rig to carry more anchor chain and therefore moor in deeper water. (The anchor chain or line must be four to five times the water depth to achieve a good mooring line catenary). In most cases, these 'upgraded' second generation semis look awful but work OK, so they can be used as a substitute for some third generation semi-submersible applications.

Third generation semis evolved from a desire to go conventionally moored into deeper water and to carry greater deck loads. Typical third generation semis are designs such as the F & G Pace Setter, the Aker H4, the GVA 5000 and the Bingo. These rigs typically have a variable deck load of over 3000 tons and can be moored in water depths of up to 600 m. As with all things in life, you do not get something for nothing when it comes to choosing a semisubmersible rig. The third generation semis will cost more than second generation semis and in a lot of cases, do not have the station keeping qualities of the second generation semis in rough seas. The reason for the latter phenomenon is simple. To gain deck load capabilities, the columns in the semisubmersibles had to get fatter. (It all comes down to displacement of the vessel when a load is applied on deck.) This has meant that third generation semis present a much less transparent profile to prevailing seas than second generation semis. Big seas can push the third generation rigs around more easily than second generation ones, hence it takes more restoring force from the mooring system to keep the rig over the well.

3.1.3 Drill ships

Drill ships are the rigs to use when drilling in calm waters which are too deep for jack-up operations. They can be used in very shallow water (±30 m) in very calm conditions but are usually used in water depths of over 100 m. Drill ships come in two main groupings: conventionally moored and dynamically positioned (DP).

Conventionally moored drill ships are rigs moored with bow and stern anchors (typically four at each end of the vessel). As such, they are restricted to water depths of up to 500 m only because there is a limit to the anchor chain or cable that they can carry. The heading that the ships are given into the prevailing seas is important since it is fixed once the anchors are set.

Dynamically positioned drill ships use a series of hull mounted propellers (thrusters) to keep them in position over the well. They hold their position by tracking geostatic satellites and transmitting interpreted data to the thrusters which fine-tune the rig position constantly. Since the rig is not anchored up, the bow can be turned into the current prevailing seas. DP drill ships will operate in water depths of 100 m up to several thousand metres and are the favoured tool for drilling wells off the Continental Shelf.

In general, all drill ships can carry huge loads when compared with semisubmersibles. This is due to their displacement type of design. (Most weight is carried below the waterline.) Modern drill ships can carry enough casing, cement and chemicals for two wells. This fact, along with their high sailing speed between locations, makes them a very attractive proposition for drilling in remote locations where there is little or no oilfield support network in place. For operations in these areas it is usual to sail the drill ship into an established oil field port to load up most of the well equipment, take the rig up to location and keep it supplied locally with fresh produce and drill crews.

When specifying a rig type, it is also necessary to specify the deck load requirements that the rig must meet. The deck load requirements are the amount of drill pipe, tools, casing, cement, chemicals, mud and fluids to be carried during the drilling operation. The greater the deck load then the less logistic support a rig needs and as mentioned earlier a drill ship can carry on board most of the equipment required to drill two wells.

Generally speaking for drill ships, the deck load is never a problem. For jack-ups and semis however, there is never enough deck load capability. As a guide, most wells of up to 4000 m can be drilled with deck loads of less then 2000 tonnes. Wells deeper than this, or in deeper water with chain moored semis, will required additional deck load capabilities.

3.2 Lifting capacity

The same criteria for lifting capacity apply equally to offshore rigs and to land rigs. Typically, offshore rigs will routinely handle larger casing sizes than land rigs and therefore will need a higher rating on their lifting capacity. Floating rigs use large seabed-placed BOP units which must be handled at surface initially by means of bridge cranes and then finally lowered to the seabed either on drillpipe or on riser. The weights involved in this operation are considerable and could in some cases exceed the normal expected drilling loads.

Furthermore, modern floating rigs are all fitted with a heave compensation system which is mounted either on the travelling block or the crown block. These will have a rated capacity of around 500 000 lbs which will be a lot less than the total lifting capacity of the rig. When specifying a floater, compensated capacity requirement should also be determined.

For jack-ups (which, as mentioned earlier in this chapter, tend to be cantilever types) the lifting capacity requirements must be satisfied by the rig in its working position. Any cantilever jack-ups must skid the derrick out on the cantilever to gain access over the well or jacket. The further out along the cantilever that the derrick must go to be above the well, the lower the rigs lifting capacity will be.

3.3 Substructure/pipe set back load capacity

This must be able to take the weight of the heaviest string of casing to be run, with a full string of pipe in the derrick. With some offshore rigs in some applications, the mechanical loads induced in this manner are less significant than the wind loads induced by high winds on a derrick which is full of pipe. This means that when specifying a rig for a hostile environment, the wind loading capabilities must be checked for the anticipated weather conditions.

3.4 BOP size, type, number and rating

The downhole pressures and temperature will dictate the rating and type of BOP to be used. Company policy or preference will dictate ram number and configuration. The BOP sizes required will depend on the hole sizes to be drilled and whether or not surface hole will require the installation of a BOP. Since on all rigs, and especially on floaters, changing BOP stacks is very time consuming, the industry trend has been towards single stack systems. Most semis and drill ships now operate with a single 18¾ inch 10 m or 15 m stack. This single large-sized BOP allows the operator to place the BOP on the wellhead once surface casing has been set and to leave it there until the well is finished.

This type of BOP will normally be specified by operators. On jack-up rigs a two stack system will usually still be needed. The BOP used on surface casing will of course be large diameter but it will not need the high pressure rating of the smaller BOP to be used, probably once 13⅜ inch casing has been set.

3.5 Pump output and rating

Once the drilling programme has defined hole configuration and hydraulics requirements, the pump requirements can be specified.

3.6 Tank capacity

Offshore rigs operate as autonomous units which can function for extended periods without logistic support. The tank system on all offshore rigs will be sufficient to cope with the drilling fluid in use up to the rig's depth capability. Additional tank capacity is required where several mud systems are being used on a well, or where completion fluid is required to be prepared and kept alongside drilling fluid. If the wells to be drilled have such requirements then the tank capacity required to achieve this must be established and specified.

3.7 Solids treatment equipment

Generally, offshore rigs require a more effective solids treatment equipment package than equivalent land wells. This is due to restricted capacity for 'dumping and diluting' on most offshore rigs. This in turn has been caused by environmental concerns which have encouraged the treatment and re-use of mud where previously it would have been dumped overboard.

If an oil-based mud is to be used then the solids treatment equipment requirements are such that only dry cuttings should be left for disposal after treatment. This will entail the use of shakers, de-sanders, mud cleaners and centrifuge systems. Furthermore, a cuttings wash system will need to be deployed to ensure that the cuttings have no oil on them prior to dumping or shipping to town for disposal.

Water-based muds do not have the exacting solids treatment equipment requirements of oil-based muds, however they will function better and more cost-effectively if the same equipment is available and used correctly.

3.8 Storage capacity

The storage capacity requirement on the rig vary according to the location and well to be drilled. Basic requirements are:

sack storage area
bulk chemical storage
drill water
diesel oil
cement
tubular goods

Since offshore rigs are purpose-built vessels, designed to drill wells of a certain depth, they will in most cases have sufficient storage capacity to cope with conventional wells. Discussions with operators who have used the rig in the past are probably the best indicator of how effective the storage capacity is in practical terms. If special operations are envisaged then the storage capacity requirements must be investigated thoroughly to ensure that the rig can accommodate the additional fluids or equipment entailed.

3.9 Accommodation

Most offshore rigs will offer between 80 and 100 beds for use by the drilling contractor, service companies and the operator. This figure will prove to be adequate for most wells. If however, extended production testing, or other labour intensive operations are planned, then it is very worthwhile considering taking a rig which has extra accommodation to suit.

Chapter 10

RIG ACCEPTANCE

The purpose of any rig acceptance is to ensure that the rig being hired is capable of carrying out the Drilling Programme safely, efficiently and cost-effectively. Most operators have in-house check-lists for rig acceptance. However, as with most lists, they tend to have grown with time, to the point where in some cases they are unwieldy and to execute the required rig audit in any depth could take longer than drilling the proposed well. It is possible to check out any working rig in a day or so. Cold stacked rigs need a longer period and are discussed separately in this chapter.

Most rigs use the same basic equipment so to get an overview of the rig the following factors must also be considered:

personnel – quality and turnover
planned and effective maintenance
inventory levels and control

Clearly, those people selected to carry out rig acceptance must be very familiar with the rig types that they are auditing. If the operator does not have such people on the staff, then it is very worthwhile considering using consultants for the job. If the operator has a Drilling Operations Policy Document, such as that discussed in chapter 3, then any rig acceptance must be done within the guidelines of this Document.

1. PREPARING FOR RIG ACCEPTANCE ON A WORKING RIG

As with any operation, good planning is paramount. We do not want to want to waste our time or that of the drilling contractor. Later in this chapter we discuss a suitable acceptance check-list. It is important that this is sent to the drilling contractor well ahead of time so that he can ensure that its needs can be met mechanically, electrically and logistically. Providing that he knows what you want to see on your acceptance visit, the onus is upon him to ensure that he has the equipment, power and personnel to arrange this. If the drilling contractor can not arrange this on the required day, then it is necessary to find out why well ahead of time, rather than have to repeat the visit later on. Whilst accepting equipment, try as far as possible to check things concurrently in order to save time.

2. PREPARING FOR RIG ACCEPTANCE ON A COLD STACKED RIG

This is a much more difficult and time-consuming process. Cold stacked rigs have generally been moth-balled using inhibitive chemicals. Providing that this has been done professionally then there should be little problem bringing the rig back into operational mode. The factors influencing the ease with which the rig can be brought back into operational mode are:

cold stacking procedures applied
inspection during cold stacking
recommissioning procedures applied

A copy of the cold stacking procedures and inspection procedures used by the drilling contractor should be sourced as soon as practicable. These documents will give an immediate impression of the attention that the operation was given.

The drilling contractor will also have a recommissioning procedure programme and a copy of this document should also be sought. If this is laid out in a flow chart form then it can be seen exactly when power will be available, load tests will be carried out and pressure testing will be done. As far as possible, we should try to work within the drilling contractor's schedule to minimise upheaval during recommissioning. Clearly, if some of these items that we want tested are not in the contractor's own list, then these have to be fitted into their programme. Exactly where they should fit in should be resolved in discussions with the drilling contractor prior to setting foot on the rig.

3. CHECKING PROCEDURE

The checking procedure falls into four main groups:

1. Personnel
2. Repair and maintenance programmes
3. Inventory carried
4. Physical check-list for acceptance

3.1 Personnel

Time should be taken to talk to rig personnel to determine their:

qualifications
experience
length of service with their present company
attitude
effectiveness

An overall impression should be gained as to ambience of the rig, by considering the following questions:

Is there pride in the rig?
Is the rig running smoothly?
How much line-supervision is evident?
What is supervisor-subordinate relationship?

A chaotic rig is usually an inefficient rig and can also be a dangerous one. A smooth operation is usually a good operation.

3.2 Repair and maintenance programmes

By inspecting these programmes, insight can be gained into the priority that is given by the contractor to maintaining his equipment. Most contractors have computerised R & M Programmes in place and these provide a quick means of establishing what has been done in the way of maintenance and what is planned. By talking to rig engineering staff an insight can be gained into the staff's confidence in the adequacy of the R & M Programme.

3.3 Inventory carried

Most drilling contractors will quote a value of inventory carried by a rig. This figure in isolation means very little to the rig's ability to cope with unexpected repairs. What is more relevant is exactly what inventory is carried.

An inspection should be made of the previous six months stores issues to correlate fast moving items with adequate stock levels. It is also worth ensuring that stock levels are being increased as a matter of course to provide spares for major PM overhaul programmes. This, together with the general condition of the store and its paperwork/software, will give a good indication of the quality of inventory handing and control.

4. CHECK-LIST FOR RIG ACCEPTANCE

The following is a guide for anyone carrying out a rig acceptance audit. With planning, it can be executed within a day or two on even the most sophisticated semi or jack-up rig. It is based on a semisubmersible rig, but with appropriate deletions and additions can be used for all rigs.

4.1 Derrick (mast) and substructure

Note the make, model and rating
Inspect for loose bolts
Inspect for bent or twisted girts and braces
Inspect for corrosion
Check block guide rails

4.2 Drawworks

Note the make, model and rating
Run the drawworks with no load to check proper function of:
 the lubricating system
 the brakes
 the catheads
 air and water supplies
 the mechanical controls
 the clutches
 Elmagco clutch, checking teeth wear
Check brake rim for wear, corrosion and grooving
Check brake blocks for wear
Run both the main drum and the sand line reel on all motor, transmission and clutch combinations
Test the foot throttle
Test the Elmagco and the Crown-o-matic
Inspect drilling line for condition and note spare length
Check ton mile programme for blockline

4.3 Rotary table

Note the make, model and rating
Test with no load at various speeds
Check amp, volts for the two mech gears
Operate the locking system (right and left)
Check and record drive gear backlash

4.4 Rotary swivel

Note the make, model and rating
Check operation
Check wash pipe integrity

4.5 Crown block, travelling blocks, hooks

Note the make, model and rating of each item
Visually check for wear, damage and misalignment
Check all grease fittings to verify proper lubrication
Run block up and down while visually observing sheaves and dollies. Listen for excessive bearing wear noises
Unlock, turn hook and relock in all possible positions
Check condition and safety equipment on all suspended sheaves
Check block trolley

4.6 Drilling instrumentation system

Test all sensors, recording and display instruments in the drill cabin

4.7 HP mud pumps

Note the make, model and number of pumps
Check proper function of lubricating system of chain cases and liner rods
Gradually vary s.p.m. from 1 to 120 in 20 s.p.m. steps. Run for 10 minutes at each step
Run on no load for 2 hours at 100 s.p.m.
Load test by pumping through the choke
Run separately each pump at 60 s.p.m. with choke open
Record pressure, amps and volts at each step
Increase discharge pressure to 2000 p.s.i. and run for 30 minutes
Increase s.p.m. from 60 to 90 to 120
Maintain 90 s.p.m. for 30 minutes and 120 s.p.m. for 60 minutes adjusting choke to maintain 2000 p.s.i.
Record Drillers' cabin indicators at each load step
Test safety devices:
chain lubricating failure
rod lubricating failure
emergency stop
A parallel test of all pumps should be carried out at 120 s.p.m. for one hour

4.8 LP mud system

Check that mud pits are clean and free from any loose material
Have them filled with seawater in such a way to check equalising valves integrity
Check dumping valves integrity
Check drillwater and oil filling valves at each pit
Check all related piping, valve by valve, at maximum system pressure attainable
Test all the mud guns
LP mud pumps: check the proper function of the various LP mud pumps, firstly with suction and discharge valves open and then with valves closed. Current and discharge pressure should be recorded. Carry out transfer test from reserve pits to active pits. Check time for above.
Run agitators for one hour each. Check current and operating temperatures
Fill tanks and traps with sea water and check valves and gate integrity
Run the pumps and test the circuits at maximum attainable pressure and with each valve combination
Record the amps and discharge pressure for each pump as for LP mud pumps
Run the shale shaker for one hour and record amps, pressures and appropriate temperatures
Run the mud cleaners, de-sanders, de-silters and centrifuge for one hour, recording amps and temperatures
Check the size, location, instrumentation and ergonomics of the trip tank

4.9 HP mud system

The HP mud piping should be tested to full working pressure valve by valve
The valves of the stand pipe manifold are to be tested to full working pressure
For each test, the pressure should be held for 5 minutes and recorded

4.10 Choke manifold, function/pressure test

Individually function test automatic choke and all manual chokes
Pressure test choke manifold as follows:

all valves upstream of chokes:
300 p.s.i. × 5 minutes
full working pressure × 10 minutes
all valves downstream of chokes:
300 p.s.i. × 5 minutes
full working pressure × 10 minutes

4.11 Drillstring and tools

Check presence and condition of these tools, cross checking with rig official equipment list.
Check all documentation and inspection certification.

5. BLOWOUT PREVENTORS

The BOP rating and certification must comply with that set down in the Drilling Policy Document as a minimum standard. To save time arrange that a test is taking place on the BOP to full working pressure as you arrive on the rig. If this is satisfactory, have the contractor perform the low pressure test then follow the inspection procedure outlined below.

The following inspections will provide an overall picture of the condition of the BOP and a summation of the conclusions enable a judgement to be made as to whether or not the equipment is suitable for the task envisaged. Since this is an equipment inspection, naturally a full pressure test at both low and high pressure will be carried out to the full rated pressure of the equipment. The elastomers used in the BOP must be checked to have a temperature rating in excess of the anticipated bottomhole temperature.

5.1 Ram type BOPs

Request that the contractor opens up one ram at your discretion then:

Inspect ram bores and cavities for overall condition
Check all bonnet faces and racetracks for marks, pitting and evidence of wash marks
Inspect condition of ram operating piston rods for marks or gouges

Check length of service from maintenance records
Record any wear to bore, etc.
Agree percentage wear with contractor to any rubber goods to be re-used (e.g., variable packers 15% worn)
Ensure equipment is available and working for correct torquing of bonnet bolts
Check wedgelock balance chambers are in good condition and function the wedgelocks
Pressure test hydraulic function of rams

5.2 Annular/spherical BOPs

Inspect and agree with drilling contractor on percent wear to elements
Record any internal wear
Pressure test hydraulic function
Inspect maintenance record as to length of service since last dismantling and re-building

5.3 BOP control pods

Inspect pipework of pods, ensure no hard pipe has been replaced by flexible hose
Have pod release mechanisms demonstrated
Pressure all SPM or slide valves to prove they seal correctly and don't leak
Function manifold and annular regulators throughout their range of pressures to check operation and freedom from leaks
Ensure recovery arms are fitted and alignment is correct
Inspect general condition of seal areas and seals
Estimate general overall condition of pods, also find out their length of service
Determine if there is a spare pod available on board

5.4 Control umbilicals and reels

Make an inspection of all umbilical terminations and ensure pod umbilical support shrouds are fitted at pods
Pressure umbilicals and look for leaks
Inspect RBQ plates
Ensure mini panel and regulators on both reels are fully functional and in good order
Run air motors on reels and check braking
Ensure there is an adequate supply of hose clamps for the water depth envisaged
Check hose lengths

5.5 Pod receptacles

Check pod receptacles for washing and overall condition
With BOP assembled check alignment of the pod and pod receptacles

5.6 Surface closing unit

Make calculations as to surface accumulator capacity requirements
Check surface accumulator pre-charge

Run both electrical and air high pressure charging pumps. Ensure correct cut out pressures are met.
Ensure all safety valves are in place
Test run mixing system
Ensure all gauges are functional
Stock check for adequate supply of soluble oil and glycol
Ensure all bolts are fitted to flame proof enclosures
Is emergency nitrogen system fitted?
Check that the pilot system has dedicated accumulators
Ensure planned maintenance has been carried out, filters, traps, oil changes, etc.
Test function control valves from satellite panels
Test low level and low pressure alarms
Have samples of control fluid sent ashore for testing
Inspect all gauges, readbacks and flow meters for correct readings
Check accumulator volume as a function of opening and closing functions

5.7 Slave panels (drillers, toolpusher's mini-panel)

Ensure all gauges give accurate readback from closing unit
All lamps should light and change correctly (done in conjunction with master panel function test)
Function flow meter
Ensure all flame proof covers at rear of panel are in place and correctly bolted. Also make sure panel is protected from elements.
Test low pressure and low level alarms

5.8 BOP hoses, pipework and accumulators

Make visual inspection of all hoses for damage and bad fittings
Pressure test all hoses, pipework and shuttle valves
Check pre-charge of stack accumulators and function stack accumulator dump

5.9 Control systems secondary support

Is ROV intervention fitted?
If intervention is fitted, list functions it can perform
Is an acoustic BOP system fitted?
If acoustic system is available, list functions

5.10 Kill and choke fail-safe valves

Determine length of service from maintenance records
Inspect previous pressure test as to proof of good service
(From above two checks, make decision as to whether valves should be dismantled to inspect gates and seats)
Pressure test hydraulic and fail-safe function
If balance chambers fitted, check oil levels

5.11 Wellhead and LMRP connectors

Check length of use in service from maintenance records
Pressure up and test hydraulic functions for leaks
Inspect gasket profile for wear and marks
Inspect dogs for wear and damage
Function connector to see smooth and proper operation
Check open and close TV indicators are in place
Record any key seating and internal wear measurements

5.12 Kill and choke lines

Check for wear and washing
Assess lines for usage, i.e. how many wells have been killed through these lines? Are they likely to be washed?
Remove any buffer flanges and inspect
If any doubt as to condition remains, have wall thickness checked ultrasonically

5.13 Kill and choke stabs and receptacles

Inspect for overall condition and absence of wash marks

5.14 Kill and choke transition hoses

Determine length of service from records
If service is long or several wells, have been killed? Have hoses removed and inspected internally?

5.15 Ball joint or flex joint

Inspect internally for wear and record same
If a ball joint is used, determine from maintenance records service life since new seals were fitted

5.16 Riser adapter flange

Determine when last NDT inspection was carried out
Have dog mechanism removed and all components inspected as to condition
Carry out internal inspection for wear and record same

5.17 BOP frame

Check all post tops are working properly
Check all guideline gates and nuts/bolts are in place and working correctly
Check all frame work for alignment

5.18 Riser

The riser supplied should be fully inspected and tested prior to acceptance. This should include any riser stored shoreside that may be later shipped offshore.

Carry out NDT inspection on all pin and boxes. This must include pins and boxes on the slip joint, the pin of the divertor, all riser handling subs and the riser adaptor flange of the LMRP.
Ensure all kill and choke lines are straight and all retaining clamps are correctly fitted. Check there is no excessive corrosion on these lines.
Dog mechanisms should be in good order. If any doubt exists, they should be fully stripped out and inspected
Inspect all kill and choke stabs and boxes for wear or wash marks. New packings should be fitted
Inspect all riser joint packings and replace any that appear squashed
The inner barrel of the slip joint should be inspected for wear, damage or key seating, any wear should be recorded
From records, find length of service of slip joint packer
Inspect condition of riser slew ring (gimbal), ensure it is free
Inspect riser gooseneck boxes for damage and wash marks. Inspect for wear or washing of gooseneck stabs.

5.19 Riser tensioners

Ensure all compressors are functional and working correctly
Check that a riser tensioner programme is in place and a proper slip and cut procedure is being followed. Make sure tonne mile records are properly kept.
Verify that all APV pressure vessels are in good order and that safety valves have been recently inspected and calibrated
Inspect control panels for overall condition and that all gauges are working
Hydraulic fluid levels in tensioners must be checked and seen to be part of a planned maintenance procedure
Hydraulic fluid samples should be sent ashore for testing
Inspect all rod seals for signs of excessive leakage
Inspect all sheaves for wear and make sure proper greasing has been carried out (no missing grease lines)
Inspect and record condition of piston rods

5.20 Drill string compensator (DSC)

We must determine that the DSC has the capacity for the work envisaged. There are also several safety hazards associated with a DSC. It is therefore imperative that all components are in good order and secure.

Inspect all hoses for damage and leaks
Check chain tensions and chains for stretch, cracked links and wear
Ensure all covers and guards are in place and correctly bolted
Function lock bar for free movement
Inspect hydraulic piston rods for damage and record condition
Ensure there is no excessive leakage from rod seals
Run block up and down to ensure dolly rollers and tracks are in good shape

Ensure safety valves are in place and hydraulic fluid levels are correct
Driller's control panel: pressure up and check for leaks. Ensure all pressure gauges are working properly. Ensure position indicator lights are working and lock bar lights change correctly.
Ensure all APVs are in good condition and the safety valves fitted have been tested and calibrated recently

5.21 Divertor system

Four main areas may give concern with a divertor system:

1. In unusual circumstances the divertor can be used to lift the complete weight of the riser and BOP. It is therefore imperative that all bolting of the divertor is throughly checked.
2. If it is known that there has been failure in use of the threaded type handling tolls, a thorough inspection of this tool is recommended.
3. Concern must be noted if the overboard lines of the divertor system contain bends and are not directly routed overboard.
4. Divertor lines must be a minimum of 12 inches ID.

Inspect the through bore of the divertor and ball joint for wear and record same
Inspect divertor packer for condition and wear, record same
Ensure all divertor seals and locking dogs are in good order and function all dogs
Check that all cap screws are in place on divertor and have their torque checked
Verify the condition of the handling tool. Ensure that inspection of the tool is recent. If of the threaded type, dismantle and inspect threads for wear.
Operate all valves in the divertor system to prove their operation and sequencing from both panels
Make an assessment of the divertor pipework layout. How straight are the lines? Will they washout if used in anger? Do they pass through enclosed spaces?
Inspect and function test the divertor control system master panel
Fully function test the driller's divertor panel and ensure all pressure gauges give the same readings as the main panel

6. SUBSEA TV

Operate below water level. Megger the subsea cable pin to pin and record the readings.

7. MARINE EQUIPMENT

7.1 Living quarters

Test all galley and laundry equipment
Check proper function of refrigerating plant, freezers and refrigerated container

7.2 Air-conditioning equipment/central heating

Check proper function of air-conditioning equipment/heating equipment

7.3 Anchor windlasses

Note number, location and rating
Note cable size, length
Function test all operations
Test sensors, recording and display instruments
Check dynamic braking
For foul weather operations check sealing capabilities at the top of the chain locker

7.4 Ballast system

Test all pumps, recording amps and pressures
Function test all hydraulic remotely controlled valves recording operating times

7.5 Bulk system

Check remote control panel operation
Pressure test the system valve by valve with air
Check all transfer and loading paths are clear by blowing through with air
Check all purge systems are working correctly
Test transfer of cement and baryte from P-tank to P-tank
Report measured time
Test transfer cement, betonite and baryte to relevant surge tanks
Report measured time

7.6 Navigation equipment

All specified equipment must be present and working

7.7 Certification

Copies of all relevant certification should be seen and noted
Any deficiencies must be investigated

7.8 Cranes

The general condition of the cranes must be checked
Any bent members in the crane boom must be noted. Rating and capacity at various boom angles must be noted.

7.9 Lifeboats

General checking should be carried out on their condition
The log book showing services and launches should be inspected

7.10 Fire-fighting equipment

A random inspection of fire-fighting equipment checking against the quoted inventory will highlight any deficiencies
A random check of dates of last service inspections will indicate how thoroughly these are being adhered to

8. ELECTRO-MECHANICAL EQUIPMENT

8.1 Load test on each generator set

Run equipment as necessary to apply test load
Adjust speed to 900 r.p.m. for synchronous operation, check and record governor settings
At no load, ¼ load, ½ load, ¾ load and full load, record all engine parameters
Run about 1 hour at full load
Observe maximum engine loading, decrease load as soon as speed is seen to decrease
(**Note:** Maximum permissible electrical load may be reached on KVA before achieving maximum KW rating)
Load step 1: 0 to ½ load, measure speed drop and time to recover with no overshoot. Modify gain and stability as necessary.
Load step 2: ½ to full load, measure speed drop and time to recover with no overshoot. Modify gain and stability as necessary.
Load step 3: Full to ½ load, measure speed increase and time to recover with no undershoot.
Check all safety devices (shutdowns and alarms) and no load
Measure temperature rise of generators

8.2 Load sharing test

Run all engine combinations at half load to check:
function of automatic synchronisation
active power sharing
Adjust load sharing equipment to correct any faults present

8.3 Motor control centres

Visual inspection
Check cleanliness of each starter/isolator cubicle, busbars, insulators, etc.
Electrical inspection
Check and record IR values of busbars.
Check and record IR values of each outgoing circuit
Function test all breakers and verify correct settings on overloads using injection test set where possible
Record all settings

8.4 Emergency generator

Running test
Check starting sequence with a simulated black out
Check alarms and shut down

8.5 Fuel oil system

Running test on all pumps recording amps and pressures
Function test all quick closing valves

8.6 Seawater cooling systems

Perform running test on all pumps recording amps and pressures

8.7 Drill water system

Perform running test on all pumps recording amps and pressures

8.8 Potable water system

Perform running tests on all pumps recording amps and pressures
Function test hydrophore system
Check proper operations of calorifier elements

8.9 Propulsion equipment

Function test cooling and lubricating systems, record pressures/amps on pumps
Check and record IR values, fans, motors and chokes
Function test controls from pilot house and control room
Function test propulsion ahead and astern at maximum achievable r.p.m. dependent on moorings and tide for minimum of two hours in each direction
Re-lubricate bearings while running
Draw oil sample from stern tube and send for analysis

8.10 Steering gear

Function test main and emergency steering systems, record amps/pressures, etc.
Relubricate rudder post bearings

8.11 SCR test

Platform visual inspection
Check cleanliness of the complete system: busbars, SCRs electronics, etc.
Electrical inspection
Check internal SCR protections: electronic overcurrent, loss of field, etc.
Check external protection and SCR alarms/shutdowns: i.e. ventilation on DC motors, MP lub oil, etc.
Check the level of current limit on drawworks, mud pumps and rotary table

8.12 400 and 600V main and 440V emergency switchboards

Make visual inspection
Check cleanliness of switchboard cubicles, busbars, insulators and all equipment in cubicles
Make electrical inspection
Check and record IR values of busbars
Check breakers, function tests in test position
Check setting of all breakers with last survey report
Check and record IR values of all outgoing circuits

8.13 Distribution boards

Visual inspection
Check cleanliness of each distribution board
Electrical inspection
Check and record IR values of each circuit
Function test all circuits, isolators, etc. and ensure correct value fuses are fitted.

Chapter 11

DRILLING OPTIMISATION

1. DRILLING OPTIMISATION AT THE PLANNING STAGE

This involves collating from offset wells and designing the well to achieve its designed objectives in the most economical manner. The main cost centres to be tackled at the planning stage are as follows:

1. Rig selection
2. Rig contract design
3. Third party service contract design
4. Wellhead selection
5. Hole design
6. Casing design
7. Directional policy and planning
8. Mud programme
9. Bit programme
10. Supervisory policies
11. Evaluation programme

In each case, the planner must ask himself whether the programmed choices are the most effective to meet the well objectives at minimum cost.

1.1 Rig selection (See also chapter 9)

In practice, the rig that you want is never available at the price you want to pay. Consequently, any choice is usually a compromise. It is the Drilling Manager's role to compare the compromises available and select the rig which can drill the well safely and at the most economic overall cost. The rig with the cheapest day rate may not represent the cheapest option. A more expensive rig with three mud pumps as against two could prove to be cheaper, if long sections of 17½ inch hole were envisaged. A rig with a top drive could save a lot of tripping time if the hole was being drilled in sticky shales or squeezing salts. There are no formulae to cover every area of the world. Each location and each hole must be judged on its own merits, all the available information collated, educated guesses made and a decision on rig selection finalised.

Prepare a check-list of rig criteria for drilling the well or wells in question. This is a particularly good idea for ease of reference if a number of rigs are to be compared. The check-list should endeavour to be as comprehensive as possible.

As was shown in chapter 5, the rig cost for an offshore well typically represents only 30 – 40 per cent of the overall drilling costs. The implications of this are that a rig which is 5 per cent more efficient than a competitor could be 12.5 per cent more expensive on day rate and still produce similar well costs.

The rig selected should have the capabilities of drilling the designed well with sufficient redundancy to handle all anticipated problems. Consider renting additional equipment if the rig is deficient in some criteria, but suitable in others. For instance, a centrifuge or extra drillpipe can be easily rented if a rig is deficient in these respects. It is not sufficient that a rig can run say 400 m of 9⅝ inch casing. Does it have sufficient pulling power to free the casing if it were to get stuck?

A detailed rig inspection should be made on any short-listed rigs (see chapter 10). The planned maintenance programmes should be inspected and items checked randomly to get an impression of general readiness for operation. The rig inspection should be as thorough as possible, with a comprehensive list of questions to be answered already prepared. Operators should have detailed rig selection guidelines to help formalise this task (see chapter 10).

The drilling equipment must be run and the inventory checked. The records of the rig's performance should be checked. Are the crews who achieved this performance still there? What turnover of crew is there? If turnover is high, then find out why. Other oil companies, who have used the rig, should be contacted for comments.

If the Drilling Programme might involve optional wells, which are deeper or higher pressured, can the rig cope with these? It is bad planning and expensive to demobilise and mobilise rigs halfway through a drilling programme. All of the foregoing points must be considered along with others prior to making the choice of rig.

1.2 Rig contract design

The operator will offer the drilling contractor the contract that he feels will protect his interests best. In practice, the operator often leaves loopholes in the contract that the contractor can easily exploit.

The normal concept of a day rate drilling contract is in fact suspect. The largest single cost in most drilling programmes is the cost of the rig. The operator would like the well to be drilled as quickly as possible and yet he puts the largest single cost centre on a day rate. The longer the contractor takes to drill the hole, the more money he makes. This is a simplistic view, but reality is not so far removed. It is not in the drilling contractor's interest to increase his efficiency too much. How often have we heard the expression 'We drilled outselves out of a job'?

This situation has encouraged the use of incentive, footage and turnkey contracts. Incentive and footage contracts are used to encourage greater efficiency on the part of the drilling contractor. Turnkey drilling hands over the entire responsibility of drilling operations to the contractor. In certain areas, only some of the responsibility can be taken on by the drilling contractor. In the UK North Sea, for example, even on a turnkey well, the operator of the licence block would be responsible for any blowout or oil spill damage which occurred on his block, even if he did not have a presence on the rig at that time.

Remember, it is the operator who writes the contract. The operator can incorporate as much or as little control as he desires in these contracts. It is probable that in the future, the oil industry will use more and more incentive-styled contracts as a means of motivating contractors towards greater efficiency.

In practice, the drilling contract should be groomed for the specific drilling programme. Local conditions should be considered. A well to be drilled in the central North Sea in June could benefit from a different rig to that required in October and November. This is due to problems with anchor handling boats in the heavy seas which are normally experienced in October. Why not consider paying an increased mobilisation cost, but only put the rig on day rate once it is anchored up at the location and ballasted down to drilling draft? In June, anchor handling is not a problem, so there is no need to deviate from the normal contract where the operator picks up the day rate from the time the rig racks its last anchor on its previous location.

A cash bonus can be offered for an accident free well or for whatever the operator desires.

1.3 Third party services contract design

As with the rig contract, it is the operator who writes these contracts. Therefore, the contract should be designed to protect the operator's interest fully, at the minimum possible cost.

All third party service tenders should bid in identical format for each service so that comparison will be easy. The use of phrases in your bid document such as 'Only charges listed in the tender will be paid, no payment will be made for any other charges', can save an operator a lot of money.

Ensure that you get the service that you want, where you want it, and that hidden costs do not appear once the operation starts. For example, if a well needs 2000 barrels of oil-based mud for a particular hole section, make sure the tender asks for that volume of mud of a required specification **on the rig**. To accept the mud onshore, transport offshore and then have to spend thousands of dollars treating it up to specification at the operator's expense, is unnecessary.

As with the rig tender, each service contract should be looked at in isolation and then with an overview. They should not just be the product of a word processor. Attempt to offer combined services as a means of reducing costs. Some companies can now offer mud logging, MWD, directional engineering and downhole equipment. Ask if they will discount further, if they get some or all of these services. If you don't ask, they probably will not offer. The ball is in our court so use it!

1.4 Wellhead selection

The costs of similar wellheads from different companies can vary greatly. The wellhead should be able to fulfil the objectives of the well and accommodate any possible programme changes. For example, on a well which is planned to have a 7 inch liner, can this liner be tied back to surface? Is the wellhead capable of handling sour gas? Will the wellhead be compatible with those already in use in your company? To choose the cheapest wellhead may not prove to be the most economical overall, since a full set of spares for it will have to be carried through the production phase. If the well is definitely going to be abandoned, then the long-term spares situation is not so important. Who, however, is so sure of the geology that he can say categorically there is no chance that the well will be a producer?

Some subsea wellhead designs require a separate trip to run and pull the annulus pack off. On a deep water well with perhaps a 7 inch tieback, this can result in considerable extra rig time, particularly if an abandonment is likely. It may be more cost-effective in these cases to run a system where the lock down rings can be removed and the pack off and hanger run simultaneously.

There are many other questions to consider:

Can the wellhead be refurbished once recovered from the well? How many times can it be re-used and at what cost? Can the wellhead be rented, are running tools supplied free of charge? Will the rig have to have a different connector? Will backup equipment be available? What backup services are offered by the wellhead company? How effective are these in practice?

Talk to other operators who have used the equipment. Most areas have regular operators' meetings; use them to gain information on other operators' experience. Of course, the salesman will tell you that the wellhead is the greatest thing since sliced bread. If three or four operators agree then he is probably correct.

Again, we must look at the situation as it applies to our well, first in isolation and then with an overview.

1.5 Hole design

Hole design is usually done backwards. The final use of the well dictates the final casing and tubing configuration. The planner then works backwards with increasing hole size and casing size as dictated by pressure regimes and lithology. Hole sizes are fairly standard but there is an increasing trend towards slim hole drilling. Slim hole drilling can be a lot faster and also reduce mud and cutting removal bills. A downhole motor can be considered here if drillpipe torque capability is a limiting factor.

One interesting recent development has been the introduction of 16 inch instead of 17½ inch hole prior to setting 13⅜ inch casing. The change might seem a small one at first glance, but the advantages the 16 inch hole can give over the 17½ inch hole are real. A better penetration rate can be achieved with a given weight of drill collars and, since the volume of cuttings generated is smaller per metre drilled, the mud treatment equipment at surface has less to handle for a given penetration rate. At shallow depths, this cutting handling capability is usually the controlling factor to the extent that quite often penetration rates of 17½ inch bits are restricted to ± 30 m/hr. A 16 inch bit can run 20 per cent higher penetration rates and still generate the same amount of cuttings. Less mud is required and consumed for the smaller hole size and less cuttings need to be removed from the site. Higher annular velocities can also be achieved for the same flow rates and better bottom hole hydraulics gained.

On the negative side, the chance of sticking the BHA is increased. How real this chance is, can only be determined by operations experience. The relationship 16 inch to 13⅜ inch is the same as 12¼ inch to 9⅝ inch, but no-one accuses the latter configuration of being liable to sticking.

Another trend being exploited in some areas is to drill 12¼ inch hole straight out of the 20 inch casing to 9⅝ inch setting depth. A contingency plan to open the hole out to 17½ inch and run 13⅜ inch casing is available only if hole problems dictate. In practice a 6 inch hole is probably the smallest hole size to be used by design. The author's practice is to use 8½ inch hole as the smallest programmed hole, with the option of setting 7 inch casing and drilling 6 inch hole should circumstances dictate this route.

1.6 Casing design

The casing bill represents a substantial part of the total well cost. The casing design should allow us to drill the well as specified and be able to handle the unexpected too. Operators' casing design policies have safety factors built in to accommodate this contingency. With a rigid casing design policy, there is not a lot the planner can do to save costs. We should look to missing out complete strings, if possible. Every well drilled in a field increases our knowledge of the area. Well Number 50 does not need the same casing design as Well Number 1. In most cases it gets it anyway!

Sometimes by moving up a grade, the costs can come down. Quite often, manufacturers get stuck with job lots of casing which are difficult for them to sell. Make sure that you get a full list of available casing and look out for bargains. Usually, casing design is based on a series of 'worst case scenarios', which may be technically sound but economically shaky.

Traditionally, for example, the 9⅝ inch is designed assuming a gas kick to surface scenario. In an area where a known pressure profile has been established, a 'limited kick' design can prove to be both technically and economically sound. Maximum dog leg tolerance whilst drilling the well will also have a bearing on casing design, when considering bending stresses through dog-legs. It is suggested, therefore, that the main design assumptions and safety factors inherent within the casing design are important decisions to be agreed at managerial level.

1.7 Directional policy and planning

If the well is directional then its implications must be studied fully. A deep kick-off will usually be more expensive than a shallow one. How much uncased hole can we stand in a kick-off section? What is the dog-leg severity we can stand? What is the minimum hole angle that we can maintain easily? How much lead angle do we build in? Do we need to penetrate the reservoir vertically? Directional wells will never make life easier for the Drilling Supervisor, but it is the planner's task to reduce the inconvenience to a minimum.

What 'cone of uncertainty' can we apply to our well path as we drill ahead? This is critical where well interference considerations exist. Running a high accuracy gyro in an adjacent well might prove expensive but allows us to drill past it with confidence using only MWD surveying. The entire survey policy must be decided at the planning stage to meet both company and regulatory body requirements. Target size and shape is a parameter that can directly affect the cost of a directional well. This should be decided at an inter-departmental level to avoid arriving at an unrealistically small target.

As with most aspects of well planning, all the variables are interactive. The BHA changes should, for example, ideally coincide with bit trips. Also, will a high inclination of severe dog-leg give rise to excessive casing wear?

1.8 Mud programme

The mud programme is based on the expected lithology, local environmental considerations and operating experience. A 'cheap' mud can be expensive and an 'expensive' mud, cheap. A cheap mud can, however, be cheap in the right hole. The trend towards exotic and expensive muds has solved a lot of hole problems in tricky wells. There is, however, no reason to use an exotic mud in an area which has been drilled quickly and safely for years with a cheaper mud.

The mud should, wherever possible, be made up with readily available chemicals. Obviously, a saltwater-based mud is easier and cheaper to run offshore than in the mountains. In environmentally sensitive areas, the use of an oil-based mud implies cutting treatment and disposal problems which cost money. Also, although exotic mud systems may largely solve our hole problems, the overall well cost could be significantly increased. The bottom line is cost and each well must be assessed on its own merits to achieve the optimal overall well cost. In addition, if large downhole losses are anticipated, then an expensive mud system could quickly become economically unviable. The properties of some exotic muds, including OBM, are very sensitive to the percentage of drilled solids in the system. If the solids control equipment on the rig is insufficient then problems could occur.

All mud parameters should be decided prior to well drilling to be kept within realistic acceptable tolerances. Running 'too tight' a mud system can incur excessive mud treatment and chemical costs. The effect of mud weight is critical on penetration rates, with most milled tooth bits in shales and similar formations. The mud weights should be kept to the minimum safe level to allow primary well control with a trip margin. Only if there are known hole problems, which can be combated by a heavier mud, should a heavy mud be programmed for a well.

The results of previous wells should be checked to see if initial pore pressure prediction was correct or if, in fact, it was over estimated. If Well Number 50 is being drilled with the same mud weight as Well 1, then either someone was very clever on Well 1 or no-one is paying much attention to the lesson of Wells 1 to 49.

1.9 Bit programme

The bit programme should look at the expected lithology and correlate that with local experience and bit records. Bit records are always suspect. No toolpusher likes to admit that he pulled a 'green bit', therefore gradings are likely to be subjective rather than objective. Similarly, the reported weight on bit may not have been run in fact, if supervision was poor. Until we have passive systems recording drilling parameters and collating them into report form, then bit records should only be regarded as a guide.

There are, however, some basic principles to apply. For milled tooth bits, a longer tooth will function better in soft formations. If hard streaks are expected, a shorter tooth can cope with them and if the formation is harder, then a shorter tooth is required. The application of basics is very subjective, however, and there is no substitute for the experience gained in similar wells in the close proximity of the planned hole. The final bit selection should be done by the on-site Drilling Supervisor based on the bits that he has pulled. However, it is up to the Well Planners to give him an initial guide and to ensure that a suitable range of bits is available for use on the rig.

The biggest single reduction in drilling costs in the 1980s was due to the introduction of PDC bits. In some formations, the correct PDC bit will not only drill faster, but it will stay in the hole longer and cause less directional problems in deviated wells. Careful selection is therefore paramount. If the well is a rank wildcat, it would seem unlikely that the correct PDC bits for the formation could be established. If however, there is offset well data, indicating that certain PDC bits will perform well, then it is almost certainly worth using them.

1.10 Supervisory policies

The supervisory policy for the rig operation can affect drilling costs dramatically. A great deal depends on the competence and work ethic of the drilling contractor. Even the best Drilling Supervisor needs to sleep at times. On tricky wells or when using dubious drilling contractors, night representatives are a necessity and not a luxury. The use of Drilling Engineers on board will be dictated by the quality of the Drilling Supervisors. Similarly, the use of Directional Engineers after the kick-off and Mud Engineers at all times, depends also on the quality of the Drilling Supervisors and/or Drilling Engineers, as well as their knowledge of that area. If the Drilling Supervisor does not know sufficient about directional control to hold angle or direction after a kick-off is completed, then possibly he should be replaced with one who does. The Directional Engineer may be costing you $700 per day to take one survey per day.

Each facet of the supervisory situation should be analysed. A bright, experienced Supervisor with a keen interest in the job can save the operator a fortune. Lack of attention to detail or lack of interest in the job will cost the operator a fortune.

1.11 Evaluation policies

To optimise on evaluation costs, we must ask ourselves the following questions:

What is the well's objective?
How can this be achieved at minimal cost?

The evaluation policy for a platform well in a known field should clearly be different from that for an exploration wildcat. More specifically we must ask ourselves:

Can MWD replace electric logging to spot formation tops?
Can we shoot sidewall samples rather than core sections?

Do the Petroleum Engineers really need the data that they are requesting, or is it merely tradition forcing them to request it?

2. DRILLING OPTIMISATION DURING OPERATIONS

This responsibility falls mainly on the Drilling Supervisor, with the Drilling Manager also being involved. The main points of interest at this stage are:

1. Drill bit selection
2. WOB
3. Rotary selection
4. Assembly selection
5. Hydraulics
6. Mud programme
7. Logistic organisation (having everything ready)

2.1 Drill bit selection

The planner will have arrived at certain conclusions regarding the type of bit required for the well. They will have done this based on available records, local experience and available bit stocks. The ultimate test of a bit's suitability is what it does downhole. Only when the bit is digging can we be sure of its suitability.

During operations the bit performance should be steadily monitored and its condition as it comes out of the hole checked carefully. Projections should be made to cost in alternative bit types. A PDC bit may have seemed expensive at the planning stage, however if a Tricone bit is not achieving its expected performance due to chip hold down, then the PDC bit may provide better economics. It is worth spending some time analysing PDC performance in similar formations, as the correct PDC bit will pay for itself many times over.

Factors such as these must be constantly monitored during drilling. The bit programme is not cast in concrete. It was at best an educated guess and the information you have about bit suitability is increasing by the hour after drilling in the formation commences.

2.2 Weight on bit

When dealing with Tricone bits there is only one weight to be run – maximum. Tricone bits tend to exhibit a straight line relationship between WOB and penetration rate (between threshold and floundering points). The hole design should be such that maximum WOBs can be run at all times. There are many ways to 'skin a cat' directionally, so pick the high WOB alternative when you can.

The WOB can easily be measured in vertical holes, however in high angle sticky holes, measurement can present a problem. If bits are coming out of the ground green after many hours and poor progress on bottom, then the problem may be hole drag. The bit was not getting sufficient weight on it to get beyond its threshold pressure.

It should be realised that most rigs do not in fact measure WOB but in fact measure loss in weight at surface. By running the largest practical diameter of drill collars, you can ensure that most of the weight loss is in fact going towards WOB. Drilling with Hevi Wate drillpipe invariably will not give you the penetration rate that a similar apparent WOB provided by large diameter drill collars will. Clearly, there are operational reasons for running smaller diameter drill collars and Hevi Wate pipe in some cases, but you should not expect to get the same progress as was possible with a similar weight of large diameter collars.

In the author's experience, the best Tricone performance will come using 10 inch drill collars in 12¼ inch hole and 6¾ inch drill collars in 8½ inch hole. The author refutes the claim that excessive annular velocity by using these configurations causes washouts since, if this were the case, the washing tendency would stop once the AV was down to a non-erosive level. In practice, this does not happen.

2.3 Rotary speed

This has an influence over drilling performance, but it is nowhere near so marked as that of WOB. The fact that it does have an influence is worth noting and rotary speeds should be varied during drill off tests. The effect of increasing rotary speed can be noted more readily in softer formations than in harder formations. It must be realised, however, that increasing rotary speeds can reduce bearing life.

Note: Whilst increasing WOB and RPM can reduce the bit life, they will usually increase penetration rate. In certain holes, a wiper trip is required every so many hours and, if the bit life can be timed to coincide with the wiper trip timing, then this will create the most economical approach to the drilling economics.

2.4 BHA selection

BHA selection should be like selecting a move at chess or playing a ball at snooker. It is easy to say what a given move or BHA will do initially, but we must ask ourselves where we will be at the end of the bit run. Ideally, any BHA change should be designed to run for the life of the bit.

Direction and angle changes should be done as smoothly as possible. For example, if we wanted to initiate a slow drop off tendency, we could utilise the fact that roller reamers do wear in abrasive formations. This factor can be used to our advantage by creating a tangent BHA with a near bit roller reamer as we run it in the hole. After a few hours, the roller reamer will wear giving a slight dropping assembly. There are a lot of 'tricks' that can be employed to achieve our aim of **maximum weight on bit at all times** when drilling with Tricone bits and **maximum hours on bit with all bits.**

In vertical hole tangent or drop off, BHAs should be selected according to local experience. Large diameter drill collars always walk less than small diameter drill collars so should be used as much as possible.

The relationship between Drilling Supervisor and Directional Engineer is critical. The Directional Engineer must know the operator's objectives **before** starting the kick-off.

2.5 Hydraulic selection

As with drill bit selection, the planners will have designed a suitable system based on known information. During drilling the well, we learn about the hole conditions and this knowledge could lead us to alter the hydraulic programme. We must look at progress, bit balling, or washing to determine whether the hydraulics are sufficient. Changing parameters should be done in a logical manner similar to a drill off test and all bits pulled must be scrutinised to see the effects of the current hydraulics programme.

2.6 Mud programme selection

As above the Planners have made an educated guess as to the correct mud programme for the hole. While drilling the well, our knowledge increases. By investigating progress, shape and size of cuttings, torque, etc. indications of the suitability of the mud system can be gained. Using MWD, we are able to monitor hole size. This information, together with all the other data that is being regularly collected, allows us to determine the suitability of the drilling fluid.

Chapter 12

DRILLING PROBLEM-SOLVING

1. PROBLEM-SOLVING MECHANICS

Drilling problem-solving can be considered analogous to fire-fighting. Whilst there are many ways to put fires out, it is always better to prevent the fire from starting in the first place. Just as with fire-fighting, fires can be prevented by separating combustible materials from ignition sources, so drilling problems can be avoided by fore-thought and awarement of potential problems. This chapter looks at problem-solving mechanics in general and then discusses the means at our disposal to prevent and tackle the two main specific problem areas in drilling which are lost circulation and stuck pipe.

Problem-solving, like most things in life, is easier if we have a plan. The basic plan to tackle problems in drilling is outlined below. This provides us with a logical thought process with which to tackle the problem.

IDENTIFY THE PROBLEM

↓

IDENTIFY THE SOURCE OF THE PROBLEM

↓

IDENTIFY THE AVAILABLE TOOLS

↓

IDENTIFY THE DESIRED COURSE OF ACTION

↓

IMPLEMENT

↓

REVIEW

The six stages of thought are broken down and discussed separately on the following pages.

1.1 Identifying the problem

To identify problems you must be looking! An old proverb says that 'There are none so blind as those who do not wish to see'. Well, those who are not looking are almost as blind! Some problems appear instantly, such as when someone drops a pipe wrench down the hole. Most problems occur initially as a trend, which increases until the operation become threatened. Examples of problems which will show trends prior to becoming serious are:

key seating
differential sticking
washouts
wear on drillstring or wellheads

All of the above can be detected by **looking**, for example:

key seating:	look at trip records
differential sticking:	look at overpulls
washouts:	look at pump pressures
wear:	look at drillstring and ditch magnet

All of these are obvious and yet on many occasions they pass unnoticed.

The Drilling Supervisor on the rig will set the standard on the rig for problem detection and identification. If he is paying no attention to these trends, then it is unrealistic to expect anyone else to be doing so.

1.2 Identifying the source of the problem

If there was not a problem yesterday then why is there a problem today? Something must have changed and the change has given the problem the catalyst it has needed to appear. The change may be a different formation, a different driller, or merely that the hole is a day older. Whatever it is, the problem should be identifiable. There may be a few potential problem sources. By listing and assessing these, the most likely source of our problem can be selected.

1.3 Identifying the available tools

Having identified the problem and its soruce, we must now consider the available alternatives. There is no point in considering displacing the hole to oil-based mud if the nearest supplier is 5000 miles away. If you are in the middle of the ocean, then tackling lost circulation with the freshwater would not be a wise choice. In any scenario, some things are working for you and some against you. We must determine how these factors affect our particular problem. We must ask ourselves if any of our available tools can be modified to suit the particular problem. Overshots can be fine-tuned for unusual fish sizes. Washing-up liquid can be pumped down hole as a lubricant or detergent. Old sacks make excellent lost circulation material if they are all that you have. Many other items can be taken out of their normal context and used in a different one very successfully.

1.4 Identifying the desired course of action

By this stage in the decision-making process we have probably defined several courses of action. The choice of which to follow will be based on two factors:

1. Which one gives us the greatest chance of success?
2. Which one exposes us to the least downside penalties?

The greatest chance of success factor is fairly obvious and is a subjective decision to be made by the Supervisor and Superintendent. The least downside factor is more difficult to quantify, since all possible eventualities must be considered prior to making a decision. The ground is full of special fishing tools invented for a specific job. In fishing operations, we should ensure that at the end of a fishing trip we are **no worse off** that we were at the beginning of the fishing trip.

1.5 Implementation

This is carrying out the course of action outlined above.

1.6 Review

This essential to analyse whether our course of action was the correct one.

Did it succeed?
If it did not succeed, then why not?
What have we done to prevent the problem recurring?
Was the problem avoidable?
If it happened again what would we do?
What 'tools' **should** we have had available?

2. LOST CIRCULATION

Lost circulation is potentially one of the most expensive problems in drilling because of the large quantities of drilling fluid that can be lost before losses are cured or reduced to an acceptable level. Lost circulation does not necessarily imply total losses and includes partial or seepage losses.

2.1 Loss zones

The range of lost circulation problems extends from shallow unconsolidated sands, to well consolidated formations that can be fractured by the hydrostatic pressures imposed upon them by the drilling mud. Furthermore, associated activities such as fluid circulation and pipe movement can cause losses. Circulation is usually lost under one or more of the following conditions:

1. Highly prorous and permeable formations such as gravel, cavernous or vugular limestones and faulted formations can allow mud particles to penetrate the formation. The degree of losses will depend on the size of the formation openings and the sealing properties of the mud in use.
2. Weak formations can be broken down by the use of excessively high mud weights, high equivalent circulating densities, pressure surges caused by running in pipe too fast, packed off stabilisers and initial circulating pressure used to break circulation in thixotripic muds.
3. Excessively high cement columns can break down weak formations.
4. Lack of care while performing leak-off/limit tests at the casing seat can result in fracturing of the formation. When using very low fluid loss muds leak-off may occur very close to the fracture propagation pressure. The tests should always be stopped as soon as any leak-off indication has occurred. Not all formations exhibit healing properties.
5. Excessive penetration rates while drilling can create a heavy mud weight in the annulus. This can cause the formation to fracture and result in losses.

If losses are experienced, they should first be quantified, analysed and then cured.

2.2 Quantification and analysis of lost circulation

1. When losses are encountered, the first priority to establish is whether or not the well is kicking.
2. Once losses are suspected, the surface equipment should be checked for leaks. On offshore rigs, check that there is not mud going into the water. On floaters run the TV camera down to visually check the riser and subsea BOP.
3. It should be established whether losses are static or dynamic. Dynamic losses are only experienced during circulation, whilst static losses occur at all times.
4. Losses occurring during drilling may be indicative of a formation change (check also the penetration rate and drilling rate), penetration of a fault, or a slight increase in mud weight and/or viscosity taking the equivalent circulating density over a critical point. In the latter case, a flow check with the pumps off will indicate whether a small reduction in mud weight and/or viscosity will suffice. Slight treatment with LCM may also be necessary.
5. It there is any doubt as to where the losses are going, consideration could be given to running a casing packer to check for losses in the casing string by pressure testing. The ditch magnet should be checked for metal cuttings.
6. An estimate can be made of the maximum weight the formation can withstand by filling up the hole with water. By measuring the volume of water taken to fill up the hole and converting this to a hydrostatic head, the formation strength can be estimated.
7. If necessary, certain wireline logging tools can be used to detect the flow of mud into a lost circulation zone (e.g. temperature survey, spinner survey etc.).

Note: For proper analysis, it is important to have as much information available as possible from the time that losses were detected (e.g. ROP, Mud wts in and out, standpipe pressure, mud properties etc.)

2.3 Curing partial lost circulation

1. As a first step, reduce the mud weight, provided it is safe to do so. Consider also reducing the ECD by reducing the PV and YP or lowering the circulation rate.
2. Lost circulation normally occurs when the pores of the formation are larger than the bridging particles contained in the mud. Sealing materials (LCM) will be available on site and can be added to the mud to cure these problems.
   When lost circulation is expected, use $^{16}/_{32}$ inch or greater nozzles to allow LCM to be pumped. If smaller nozzles are used, employ a circulation sub. If it is considered necessary to pump LCM pills through the bit, the nozzles can be blown out using Schlumberger primer cord.
3. A 25–50 bbls pill containing a mixture of lost circulation materials at a concentration of between 5–30 ppb, depending upon the degree of losses, can often cure lost circulation. The pill is preferably spotted where the losses occur and allowed to stand there for a short period. If the losses are not cured, a second pill can be used, or if there is a large section to be drilled, then lost circulation material can be added to the total mud system at about 2 ppb. Bypass the shakers when LCM plugs the screens.
4. If losses are the result of a kick, then the kick should be controlled first by one of two means:
   (a) Displacing a heavier mud in the open hole below the lost circulation zone and decreasing the mud weight above it.
   (b) Setting a baryte plug (max. wt. 21 ppg) in open hole below the lost circulation zone.
5. If a well kicks as a result of lost circulation then the lost circulation has to be cured first before the kicking formation can be controlled. This can be done by pumping LCM pills either down the annulus or DP.
6. If none of the various LCM pills are successful, drilling without returns may be considered.
7. After having experienced lost circulation problems, setting casing as soon as a non-permeable zone is encountered should be considered. If it is known that pore pressure gradients deeper down the hole do not increase, then it may not be necessary to set casing immediately. Never pull out when circulation is lost and the level cannot be seen until the situation has been fully accessed. The main danger in doing so is the possibility of swabbing the well in. With no level to be seen, it is impossible to detect swabbing, except by watching the weight indicator.

2.4 Lost circulation pills

1. Slight losses can often be cured by:

Mica fine	3–5 ppb
Mica course	3–5 ppb

2. Severe losses generally require a greater range of particle sizes and the following may be used, depending on severity:

Mica fine	3–10 ppb
Mica medium	3–10 ppb
Mica course	3–10 ppb
Cellophane	3–10 ppb
Nutplug	3–10 ppb
Mud fibre	3–10 ppb

The amount of LCM that can be added to a mud depends very much on the mud properties. High gels and YP will help keep the material in suspension. If too much material is added, then the excess will float on top. For this reason, add the material in stages, for example 5, 10, 15, 20, 25, 30, 35, 40 ppb. Stop after it is observed that the material cannot be held in suspension. If the LCM pill is left to soak for a period of time, you may find that more material can be added. It is useful to have a viscofier, such as CMC, available to thicken the mud before adding LCM material.

2.5 Gunk squeeze

If losses are particularly bad, a gunk squeeze can be tried. This method utilises the swelling property of bentonite in the presence of water and is done as follows:

1. The cement pump and lines are thoroughly cleaned with diesel oil to remove all traces of water.
2. A diesel spacer is then pumped down the drillstring with the bit or tail pipe just above the lost circulation zone.
3. 300 ppb bentonite in diesel oil is pumped down in a 20 bbls pill.
4. Follow this gunk mixture with diesel oil spacer.
5. Displace gunk mixture to the bit with mud.
6. Pump simultaneously down the drillpipe and the annulus and apply a squeeze pressure of 100–300 p.s.i.

While the bentonite is pumpable when mixed in diesel, as soon as it comes into contact with water it forms a highly viscous putty like material which can plug off the formation. This method is particularly effective if used in holes where considerable amounts of water have been lost downhole prior to attempting the gunk squeeze.

2.6 Use of cement plugs

When losses in excess of 100 bbl/hr are encountered, or when operational circumstances necessitates, cement plugs should be used to cure losses. Light weight cement should be used such as 1.67 SG 2 per cent bentonite slurry. The cement plugs should be placed over the loss zone and drilled our prior to full hardening to prevent the bit from wandering off them into the formation which is probably softer than the cement when it has fully hardened.

2.7 Recommended stocks of LCM

It is recommended to have the following minimum stocks of LCM at the rigsite at all times:

1 pallet mica (fine)
1 pallet mica (medium)
1 pallet mica (coarse)
1 pallet viscofier (to make the pill viscous before adding LCM)

In areas of known losses, these stocks must be increased significantly.

2.8 Bit nozzle removal

Under certain conditions, it is desirable to remove the nozzles from a bit downhole to enable free passage of plugging/lost circulation materials. Nozzles, whether retained by circlips or nails, can be removed using explosives. In order to use sufficient primer cord the minimum ID, in the drillstring should be $2^{1}/_{16}$ inch. The charge is made up of a firing head of ⅜ inch mild steel bar approximately 7 ft long with a 1 inch OD, ×2 inch long bull nose on the bottom (the basic string shot tool). One strand of primer cord is used for detonation with twelve lengths of 16 inch long primer cord laid along the bottom part of the bar. Wrap approximately 3 ft primer cord around the 12 lengths at approximately 9 inch from the bottom of the tool. Finally, wrap plastic tape over the cord to hold the charge in place. 80 grain primer cord is ideal. The maximum OD of the charge should be 1¾ inch. String connections adjacent to the bit can be slackened by this operation, therefore, care must be exercised when tripping out afterwards.

2.9 Total losses

Sometimes losses are not curable. If you drilled into a coal mine or subway, it would be approaching the impossible ever to fill the hole effectively. Some formations are analogous to this, it is just not feasible to cure the losses. This is not common in practice and tends to happen only in surface hole, fractured limestones and when drilling in caves and mine workings. At this time, it is a problem which must be tackled. The first question that we must ask ourselves after checking that the well is not flowing is, 'What is the hole going to do?' Having discovered the loss zone and lost the fluid head of mud which provided the primary well control in the borehole, any hydrocarbons would make their presence felt very quickly. At the same time unconsolidated formations higher up the borehole may fall in as the pressure differential across the borehole wall is increased with the loss of hydrostatic head of the mud. Consequently, our two main concerns can be addressed by the questions:

Will the well kick?
Will the hole collapse?

As mentioned before, if the well is going to kick with no fluid in the hole, then we should know about it fairly quickly. A gas influx will almost certainly flow from the open hole and require closing in at the BOPs. Oil and water kicks could come to surface, but could in fact just produce into the loss zone. In this case, the well might be producing with little or no clear indication of this at surface. If this is the case, then we need to be aware of it to ensure that any future casing programme and cementations take this into consideration.

To check the likelihood of this eventuality, the geologist's records of the well to date should be checked, along with drilling data such as background gas, oil shows and salinity. If none of the formations drilled to date had gas, oil, or water shows and furthermore, if they were non-porous and impermeable, then there is little or no risk that any formation is producing. If there is a strong likelihood, however, that there is an exposed productive reservoir, then it must be assumed that this will produce under total loss hole conditions. Quite often with the drawdown created across the well bore interface during total loss downhole conditions, the productive formation will bridge itself off and seal off. This phenomenon, although forming the basis of a lot of wild well control planning, can never be relied on. If there is oil present in the well bore, it should be evident on the drillstring as and when it is tripped out of the hole.

Whith total losses, we face the alternatives of curing losses or drilling ahead blind. Our initial attempt should be to cure losses using the methods mentioned earlier in this section and then, in the event of their failing, pouring in light weight or thixotropic cements to cure the losses. If formations have been fractured and unconsolidated since spudding the well or since the last casing was set, then it is worth considering drilling ahead. Oil and gas in a reservoir needs a 'cap rock' to retain it and prevent it from migrating to surface over the millions of years during which it was forming. If we have not drilled through any formation which could be considered to be a cap rock or a seal, then it is extremely unlikely that any hydrocarbons will be encountered until we do actually reach a formation which could be a cap rock. Casing should be set in the first firm formation reached.

3. STUCK PIPE

3.1 Causes of stuck pipe

Pipe gets stuck downhole for a variety of reasons. These reasons can be broken down into two main groups:

1. Differential sticking
2. Mechanical sticking

3.1.1 Differential sticking

Since whilst drilling, the hydrostatic head exerted by the mud column is greater than the pressure of the formation fluids in permeable formations, mud filtrate flows from the well into the formation leaving a filter cake on the wall of the bore hole. A pressure differential will exist across the filter cake, equal to the difference in the hydrostatic head of the mud column and the formation. The ECD created whilst pumping, increases this pressure differential.

When the drillstring comes in contact with the filter cake, any part of the pipe which becomes embedded in the cake, will be subject to a lower pressure than the part which remains wholly in the well. If the pressure difference is high enough and acts over a sufficiently large area, the pipe may become stuck. The force required to pull differentially stuck pipe free, depends upon:

1. The pressure differential between the well bore and formation. This differential creates a force pushing the pipe against the side of the hole, adding to any lateral forces present due to hole deviation.
2. The surface area of pipe embedded in the wall cake. The thicker the cake or the larger the pipe diameter, the greater this area is likely to be.
3. The coefficient of friction between the pipe and the wall cake. This will depend on the mud in use and the formations which have been drilled through.
4. Differential sticking should be thought of as analogous to surface tension on the surface of liquids. The pipe is essentially stuck in an elastic medium and this situation influences our freeing methodology.

3.1.2 Mechanical sticking

Mechanical sticking is usually caused by one or some of the following:

1. Junk in the hole
2. Insufficient hole cleaning
3. Unstable formations
4. Key seating
5. Poor drilling practices

1. Junk in the hole

Most junk in holes comes in from the surface. Pipe wrenches, slip dies and tong jaws are commonly dropped downhole inadvertently. This type of junk will generally stick the drillstring as it is being pulled up, either on trips, or whilst making connections. Junk, can however, be created downhole by tool failure in such tools as rebel tools, under-reamers and Tricone rock bits.

2. Insufficient hole cleaning

If cuttings are not effectively removed from the well, then when circulation stops, they can settle around the drillstring, causing the hole to pack off and the pipe to become stuck.

3. Unstable formations

The stability of formations encountered downhole varies greatly. Some formations are inert and are no problem to drill through, whilst others such as shales and salts, can cause extensive problems. Troublesome shales fall into two groups, swelling or sloughing.

Swelling shales react to water and as their name indicates will swell with time. The speed at which swelling takes place will depend on the type of drilling fluid in use and the shale type itself. Swelling shales can be detected during drilling as overpulls, as larger diameter drilling components are pulled through swelled sections. The cuttings circulated up from drilling through these sections will also give an indication of the type of shale at the bottom of the hole, since they will quite often yield on the way out of the hole to create clayballs at surface.

Sloughing shales degrade downhole by having fragments falling off into the well bore in the form of cavings. Cavings have a totally different appearance from cuttings and therefore can be spotted at the shale shaker. As the mud weight is increased, more filtrate can be forced into the shales which can increase the amount of cavings. It is also possible that increasing the mud weight can reduce the problem by bringing hydrostatic mud pressure closer to formation pressures in the case of overpressured shales. Field experience alone will indicate how the local sloughing shales will react.

Most salt sections can be drilled without problems, providing that super saturated salt or oil-based mud is used. However, the plastic nature of some salts will inevitably cause hole problems. As the hole is drilled through, the salt section stresses are relieved and the salt flows plastically into the borehole. In some areas, this requires special drilling practices, such as regular reaming, to allow the well to be drilled ahead.

4. Key seating

This is caused by the rotation of the tool joints and drillstring against the formation. It typically only manifests itself at abrupt angle changes in the well. The extent of key seating is dependent upon the abruptness of the hole dog-leg, the softness of the formation there, the amount of tension in the drillstring which acts to force the string into the wall of the hole and the rotating hours.

Key seating is detected initially on trips as a blip on the drilling recorder chart. This blip indicates a slight overpull as the BHA is pulled into the key seat. At the first sign of this, the key seat should be wiped either using a proprietary key seat wiper or a stiff stabilised BHA. Once detected, the key seat area should be tripped through with extreme care, as even when a key seat has been effectively wiped out, a new key seat can develop during subsequent drilling operations.

5. Poor drilling practices

Whenever anything changes in a well, the Drilling Supervisor must consider the effects that these changes might have. When changing from a limber BHA to a stiff BHA, care must be taken, as the string is tripped into the well, to prevent it from sticking. If reaming is being carried out, it should be done gingerly so that the BHA is not 'screwed into' the formation. Similarly, if an undergauge BHA has been pulled, care must be taken when running a full gauge BHA on the next trip in.

When pulling out of the hole, the driller should be advised as to what overpulls are permitted, prior to lowering the string again, or if necessary back reaming out of the hole.

3.2 Prevention of stuck pipe

The prevention process begins as the well is programmed and continues whilst the well is being drilled.

3.2.1 Prevention of stuck pipe at the programming stage

When writing the Drilling Programme, each formation and situation must be assessed with its potential as a source of stuck pipe problems. The reaction of formations in offset wells must be studied and responded to in the Drilling Programme.

Correct mud system selection is paramount for a successful well. The use of oil-based muds in troublesome shales has, in some cases, eliminated the problems. In other formations, oil-based muds can create as many problems as they solve (see chapter 7). Keep mud weights to a working minimum. Programme the use of Hi Vis or Lo Vis pills for hole cleaning, if problems are anticipated.

More and more rigs are coming equipped with top drive systems. These have taken a lot of the excitement out of back reaming and should be specified if other methods of combating squeezing salts or heaving shales have proved to be ineffective in the past. Troublesome formations sometimes only rear their ugly heads after a finite open hole time. This is especially true in directional wells. In this type of formation, typically shales, the likelihood of expected hole problems can be thought of as being proportional to the square of the time in days that the hole is open.

With these formations, it simply means that if you can put the well in fast, then you will see few problems. There are limits to how fast wells can be drilled, however, and in some cases it may be necessary to plan an extra casing string just to consolidate the hole prior to its beginning to cause problems. Do not programme mud motor kick-offs or any other activity which involves keeping the drillstring stationary in hole sections which are known or suspected to be potentially troublesome. In extreme conditions, consider using MDW rather than electric logs for some evaluation purposes, since they are less likely to get stuck.

3.2.2 Prevention of stuck pipe on the rig

The Drillers, Mud Engineers and Mud Loggers must be kept informed of potential stuck pipe problems and how to respond to given situations in order to preserve the hole's integrity.

Drilling instrumentation must be in perfect working order and the mud logging unit running according to specifications.

Be aware of and note on reports any drags, torque fill on trips, and cavings.

The Drillers, Mud Engineers and Mud Loggers must report any and all detected applicable trends to the Drilling Supervisor, who in turn must consider the appropriate response.

When running in the hole do so in such a manner that at no point is more than the BHA weight slacked off at a tight spot, without calling the Drilling Supervisor.

Perform wiper trips in response to hole conditions, not according to any pre-set levels. The author has, on occasions, wiped troublesome holes every four hours to keep the string free. On other occasions a wiper trip does nothing.

Keep the BHA as small as is practical to supply effective weight on bit.

Keep the BHA well stabilised and utilise an undergauge stabiliser on the top of the BHA to facilitate back reaming.

Keep the drillstring moving as much as possible when in open hole.

Overlay the trip charts on successive trips to highlight potential key seats and squeezing formations.

Integrate this overlay with known dog-legs in the well and highlight these depths on each trip.

Record all depths of sticky hole when drilling. Study the mud log as cuttings are returned to surface and trouble-shoot potential trip problems.

The operator's Drilling Supervisor should be on the rig floor for every trip out through newly drilled hole and known hydrocarbon bearing formations.

Sufficient circulation rate should be used in all hole sections to clean the hole.

Prior to trips out, the mud must be conditioned to bring solids content down to designed levels and to ensure that all drilled solids are out of the well. Check at all times that the hole is being cleared effectively by the mud.

The BHA should be inspected as it comes through the rotary for clay deposition, especially on stabiliser blades. In most water-based muds, this deposition can be eliminated by dowsing the system with a couple of barrels of soap or detergent during the pre-trip circulation.

An effective drilling jar should be used in every drilling BHA situation such that it will not be run at neutral weight during any envisaged drilling scenario.

A pipe wiper should be used at all times to cover the hole to prevent junk falling down through the rotary table.

Always measure and check all equipment being run in the hole. Always use the correct gauge ring for each tool.

After pulling a worn BHA always ream carefully back to bottom.

When running a stiff BHA following a flexible BHA run, always ream carefully back to bottom.

3.3 Freeing stuck pipe

When the drillstring becomes stuck, immediate action will prove to be most effective in most cases. Prior to taking this action, the cause of the problem must be determined. This is achieved by considering the circumstances leading up to getting stuck and relating this to well conditions and history. Pipe will be stuck, either mechanically, or differentially. Sometimes, mechanically stuck pipe will end up differentially stuck too, but even in the worst hole, this will take time.

3.3.1 Identifying the cause of stuck pipe

At the time of getting stuck we have to consider the following:

Can we pull the string?
Can we lower the string?
Can we rotate the string?
Can we circulate?
What operation was being carried out at the time of sticking the string?
What were torque and drag trends prior to becoming stuck?
What incidents occurred in nearby wells of the same depth?
What is the BHA design?
What formation is opposite the BHA?
Are there any other salient details?
  junk dropped downhole
  loss in circulating pressure
  excessive cuttings over shakers
  no cuttings over shakers.

By considering these details, the likely cause can be determined and this may indicate whether the pipe is stuck mechanically or differentially.

3.3.2 Freeing mechanically stuck pipe

If the pipe was in motion at the time of getting stuck, then the first attempt to free it should be in reverse to this motion. For example, if the string has been pulled into a tight spot, then it should be worked or jarred downwards. Conversely, if the string has been run into a tight spot it should be worked or jarred upwards.

If the mechanical sticking is due to inadequate hole cleaning, then pump rate should be increased to the maximum operationally permissible in an attempt to assist hole cleaning. High circulation rates can also encourage washing out and consequently enlarging the hole in some cases.

If a drilling jar is in the string, the jarring should be carried out in the opposite direction to which the string became stuck. If the string was pulled into a key seat, jar downwards and if the stabiliser jammed in the formation whilst reaming down, jar upwards. If a drilling jar is not in the string, then the pipe should be worked in the same direction as above to the safe working limit of the drillstring.

By looking at the formation which is sticking, an indication of which chemical approaches might be successful can be arrived at. If the string is stuck in carbonates, then a slug of 2 per cent HCl can be used to eat into the formation and release it.

When stuck in salt, freshwater will dissolve the formation and free the pipe in most cases.

3.3.3 Freeing differentially stuck pipe

This can be done in three ways:

1. Working pipe
2. By reducing hydrostatic pressure
3. Chemically

1. Working pipe

In most cases of differential sticking, the pipe will be stuck with the bit off bottom. Because of this, working pipe can be done either up or down. Working pipe and jarring is rarely effective in freeing differentially stuck pipe, but it is easy to do and worth a try in most cases.

2. Reducing hydrostatic head

Since the sticking force is proportional to the differential pressure existing between the mud, hydrostatic pressure and the formation pressure, then by reducing the differential pressure, the sticking force can be reduced. This reduction of hydrostatic head can be achieved by lightening the mud or circulating to a lighter fluid such as water or diesel. Alternatively, the pipe can be U-tubed. All of these methods must be done whilst still retaining full primary hydrostatic control on the formation. These methods have mixed success rates as quite often the filter cake remains in place between the pipe and formation sticking the pipe, even with reduced differential pressure.

3. Chemically

In most cases of differential sticking, the pipe will ultimately be freed chemically (if at all). To do this a pill of chemical agent which attacks wall cake is spotted at the stuck point. This is usually mixed with diesel oil and weighted to the same weight or slightly heavier than the mud weight in use. Chemical pills take time to work and 24–36 hours should be allowed for this prior to giving up. Enough pill should be made so that the pill can be moved every hour or so. In practice, this means making a pill of approximately twice the volume required to fill the drill collar/open hole annulus and leaving half of it in the drillstring. The pill is then moved by pumping a few strokes with the mud pumps at regular intervals. If water-based mud is in use, then a very effective method of freeing differentially stuck pipe is to displace the hole to oil-based mud. As with all chemical methods, this takes time to free the pipe and 24–36 hours is a typical delay before the pipe is free.

Chapter 13

LAND DRILLING PROJECT MANAGEMENT

Anyone who thinks that drilling on land is easier than drilling offshore, has probably never tried it. The planning and operational challenges in land drilling are different from those in offshore drilling but no less significant. Offshore rigs tend to be totally self-contained units with proven working systems and logistic support networks. Land rigs have to start from scratch on most locations and each new location presents its own particular environmental service or logistic problem.

Offshore or onshore, planning and preparation are the key factors in ensuring a safe, cost-effective and efficient drilling operation. Planning a land well can be broken down into three groups:

1. Pre-planning
2. Detailed planning
3. Drilling well

Each of these groups can be broken down further into a total of 26 sub-groups. This is laid out below and Figure 1 is a flow chart which shows the duration of each component with respect to a North-west European land well. Following on, each sub-group is discussed with its implications to the operation.

1. PLANNING A LAND WELL

Pre-planning

1. Selection of drilling prospect
2. Group agreement
3. Construct planning bar chart

Detailed planning

4. Preliminary site reconnaissance
5. Discussions with local authorities
6. Initial discussions with landowners
7. Discussions with mineral rights holders

8. Site selection
9. Site surveys
10. Site plan
11. Obtain agreement in principle with landowner
12. Submit planning application for well
13. Contact with local groups
14 Sign up site contracts
15. Prepare Drilling Programme and AFE
16. Obtain AFE approval
17. Emergency Procedures Manual
18. Submit Drilling Programme to government authorities

Drilling well

19. Select drilling contractor
20. Procure materials
21. Select services
22. Build site
23 Mobilise rig
24. Drill well
25. Test/suspend/complete/abandon
26. Restore site

ACTIVITY	TIME IN MONTHS													
PRE-PLANNING	1	2	3	4	5.	6	7	8	9	10	11	12	13	14
1. SELECTION OF DRILLING PROSPECT														
2. GROUP AGREEMENT	X													
3. CONSTRUCT PLANNING BAR CHART	X													
DETAILED PLANNING														
4. PLANNING SITE RECONNAISSANCE	X													
5. DISC. WITH LOCAL AUTHORITIES	X													
6. DISC. WITH LAND OWNERS	X	XX												
7. DISC. WITH MIN. RIGHTS OWNERS	X	XX												
8. SITE SELECTION		XX												
9. SITE SURVEYS			XX											
10.SITE PLAN			XX											
11.AGREEMENT WITH LANDOWNER			XX											
12.SUBMIT PLANNING APPLICATION			OO	OO	OO	OO	XX							
13.CONTACT LOCAL GROUPS				XX	XX									
14.SIGN UP SITE CONTRACTS				OO	OO	OO	XX							
15.PREPARE DRILL PROG. AND A.F.E.				OO	OO	OO	XX							
16.OBTAIN A.F.E. APPROVAL							XX							
17.WRITE EMG. PROCEDURES MANUAL							XX							
18.SUBMIT DRILL PROG. TO D.O.E.				OO	OO	OO	XX							
DRILLING WELL														
19.SELECT DRILLING CONTRACTOR								XX	XX					
20.PROCURE MATERIALS									XX					
21.SELECT THIRD PARTY SERVICES									XX					
22.BUILD SITE									XX	XX	XX			
23.MOBILISE RIG											X			
24.DRILL WELL												XX	X	
25.TEST/SUSPEND/COMPLETE/ABANDON													X	
26.RESTORE SITE														XX

Figure 1 Norwell flow chart for UK Land drilling

1.1 Pre-planning

1. Selection of drilling prospect

This is done in-house by OPCO Explorationists.

2. Group agreement to drill an exploration well

This is done in-house by OPCO with input from lease partners. At this time it is necessary to create a premliminary well programme and order of magnitude cost estimate to obtain budgetary approval.

At this stage of the project, a local agent should be appointed. Overseas, this would tend to be a well-connected consultant who could maintain government and local political contacts. In the UK, it would be an estate/planning agent with good planning authority contacts. Such people need not be expensive and have the enormous benefit of local expertise. The alternative of having OPCO employees in-situ is normally ridiculously expensive and is not recommended.

3. Construct planning bar chart for well

At this stage, a provisional activity bar chart plan for the well should be constructed. The attached bar chart is shown merely as an example of the time scales involved. It is fairly typical of European land wells. However, each component part could vary substantially in different areas. Critical activities will nearly always be site selection, planning permission and choice of rig.

Gaining authorities' and landownders' permission can be a major stumbling block. In the UK, for example, it is generally relatively easy to obtain a site and local planning permission in any area of the Midlands, the North and Scotland. Opposition from local environmental groups is especially noted in the South and has held up many drilling projects. Environmental concerns tend to be a problem almost everywhere in Europe and good planning and liaison are required with the local authorities from day one, to avoid unnecessary expense. Outside of Europe, environmental concerns can be less, but often accessibility, water supply and technical services become major factors in site selection.

Rig selection can have a major impact on costs. In general, one should select the cheapest rig that can do the job. In Portugal, for example, where there is no oil field infrastructure, on a Norwell drilling project it proved possible to beef up the equipment of a Water Well drilling contractor. This simple expedient, which was achieved by forming a joint venture between the contractor and a UK drilling company, saved over $180 000 on the well cost compared to mobilising on oil field rig from Spain or the UK.

1.2 Detailed planning

4. Preliminary site reconnaissance

Once a bottomhole target has been selected by the Explorationists, then it is necessary to make an 'in situ' inspection to determine obvious merits and demerits of the area. The site should be as small as practical. The rig selection should be made on the real requirements of the hole and not on a 'bigger is better' criterion. Natural screening should be used whenever possible and the absolute minimum of road laying and utility laying should be achieved by positioning the site correctly. Good access to the site should be assured by paying attention to local low bridges and overhead power cables. Ease of providing communications should be considered and the implications of a disaster at the site should be considered. Any problems caused by the existence of mineral rights holders in the area should also be identified.

A short list of suitable surface locations should be drawn up and the implications of drilling directional wells considered and costed out, in the event that a vertical well prospect proves unfeasible.

5. Discussions with local authorities

The importance of local authorities as opposed to central government varies enormously worldwide. In Northern Europe, it is essential to establish good relations as early as possible, in order to avoid delays. In other parts of the world, the local authorities often have less power and are more interested in promoting local economic activity. Good relations may be easier to establish and require less effort to maintain. In any event, local authorities can define problem areas and local requirements. It is essential to keep them informed as much as possible, so that they do not feel shut out. Their co-operation is vital. At the very least, they can normally provide advice on site ownership and mineral rights.

6. Initial discussions with landowner

The landowners are most easily tracked down by using a local agent. It is important to have alternatives, since a single source will inflate the price. Whether it is best to purchase or lease the land depends entirely on the project under consideration.

7. Discussions with mineral rights holders

Mineral rights holder's agreement is usually required before the governmental energy authorities can approve the Drilling Programme if mineral measures will be drilled through. Normally, the mineral rights holder will list requirements to be built into the Drilling Programme such as:

- only GRP casing can be used within 45 m of coal seams or potential mineral workings
- one month advance warning of spud
- one day advance warning of logging activities
- surface and intermediate casing strings to be cemented to surface
- production casing strings to be cemented minimum 100 m inside next shoe
- written mineral rights holder's consent prior to well abandonment

8. Site selection

Final site selection will normally require input from the Drilling Superintendent, the Explorationist and the local agent. The following matters should be considered in arriving at a decision.

- access to site
- natural screening

existing utilities
proximity to habitation
environmental consideration
size and shape of site
proximity to airports
proximity to military activities
proximity to roads/railways
cost including site purchase (or lease), site preparation, influence on well cost (deviation)

With the information accumulated, the sites should be listed in order of desirability and serious negotiations begun to secure use and access.

9. Site surveys

Once a site is selected, the following surveys may be required:

environmental survey
topo survey
access survey
noise limitation survey

10. Site plan

The site plan should meet the following criteria:

be large enough to accommodate the rig and equipment
provide easy access for solids and waste removal
allow the placement of production testing equipment without violating zonal safety
provide sufficient solids settling pits
provide a safe location for a flare line from the BPM
allow off-road car parking
have sufficient room to turn trucks around on site

To solve all this cost-effectively, it must also be as small as possible. Essentially, we are faced with a compromise between size and cost. In most cases, the overall well cost will be reduced, if the maximum size of cuttings ditch is used. Trucks shuttling backward and forwards to the rig site to remove fluids and cuttings are an operational inconvenience and an added expense. The cutting ditch should be big enough to accommodate at least twice the calculated volume of generated drill cuttings. A large size is most convenient and typically some larger operators seek to have a plot of 100 m × 100 m or greater. However, if this size introduces a serious cost penalty, it is possible to revise the rig layout and schedule stock deliveries to reduce the site size requirements. Approximate minimum site size requirements for conventional rigs would be as follows:

1000 m rated rig	55 m × 40 m
2000 m rated rig	65 m × 50 m
3000 m rated rig	90 m × 65 m

It will also be necessary at this stage to decide on the requirements for site sealing and drainage. In the UK, even in environmentally sensitive areas, a membrane (plastic) seal is considered adequate, together with a bunding and collection of surface water. In Holland, for example, concrete sites with bunding and inner and outer drainage systems are required and sites typically cost $400 000 or more.

11. Obtain agreement in principle with landowner

In environmentally sensitive areas, a legally binding agreement may be required at this stage to stop the landowner reneging.

12. Submit planning application for well

This activity is one that is entirely defined by local regulations. The following comments can therefore only be considered as a guide.

In urban areas site construction work hours should be limited to 08.00–19.00 weekdays and 08.00–13.00 Saturday with no work on Sundays.
In urban areas, conductor driving should be avoided if at all possible.
All top soil removed whilst preparing site must be stored for site recovery work.
Site access must be clearly marked on plans and heavy traffic to and from the site restricted to normal working hours except in an emergency.
Restrictions may be placed on the hours during which hydrocarbons can be flared off.
Only two residential Porta-cabins will usually be allowed on site.
Rig lighting should be such as to provide minimal glare 500 m distant.
Rig trucks will need to be cleaned prior to leaving rig site to ensure that no mud/clay deposits are left on the public road system.
Additional road signs should be erected to ensure that site traffic takes the route of minimum local disruption and maximum ease of access.
Policy on removal of waste water, effluent and cuttings should be presented.
Two months is probably the minimal period for obtaining planning permission in the UK.
Announce drilling plans to news media if it is company policy to do so, publish a news statement about the well or hold a press conference.

13. Contact with local groups

It is at this stage, in Europe, that local groups may start to object to drilling plans. Again, this is a problem which is encountered on very few wells, even in the UK. It is recommended that the OPCO do everything possible to avoid environmentally sensitive areas. Should this prove impossible, the main groups to deal with will be:

parish council
local residents
environmental groups

Allaying their concerns is possible, but expensive, in terms of manpower and direct cost. A public relations campaign can only be mounted effectively with the assistance of a local PR firm. A low-key exhibition can be much more effective than public information meetings, which can be confrontational and unsuccessful.

14. Sign up site contracts

Finalise access arrangements with landowner. Contracts must be signed at this time. Finalise mineral rights holder's approval (if applicable). Again written approval is required. Obtain drilling permission from water company (if applicable). The local water company will normally wish to ensure that there is no contamination of their water. The proposed well must be studied from the land drainage or water reservoir contamination point-of-view and the water company satisfied that their resources are not threatened. An official application should be made at this time, although the potential implications should have been considered much earlier to highlight and overcome any problem.

15. Prepare Drilling Programme and AFE

At this stage, the first draft of a detailed Drilling Programme should be prepared and quotes solicited to allow an accurate AFE to be prepared for partner approval.

Mud policy should be as cheap and simple as possible. One system from top to bottom if possible. Investigate local supply of chemicals, especially bentonite.

Casing design should be approached with no preconceived ideas. Meet well requirements and then cost alternative ways to reach the objectives. Bear in mind that larger holes produce more cuttings, need more mud to fill them and need more cement to case off or plug off.

Cement design should be as cheap and simple as possible. Ideally this should be the cheapest, most readily available cement mixed with the local water supply. Any additives cost money and must be justified. Typically, oil well cement and equipment is expensive and local readymix equipment may be a cost-effective solution, especially at shallow depths.

Slimhole drilling should be checked by a feasibility study. The costs and availability of unusual bit and casing systems must be offset against savings in mud, cement and cutting removal.

16. Obtain AFE approval from partners

This will normally take from one to four weeks.

17. Emergency Contingency Manual

This should be kept as simple as possible. Standard procedures for dealing with well kicks and safety should be dealt with in the OPCO Operations Manual. The ECM should deal with the procedures to be followed in case of a blowout and/or fire and define the relationship with the local emergency services (Fire, Police and Ambulance).

18. Submit Drilling Programme to governmental energy authorities
Government authorities are tending to require more and more technical detail on the well. It is best to submit and explain drilling plans as soon as possible, to allow the government authorities time to study and comment on your Drilling Programme.

1.3 Drilling well

19. Select drilling contractor
The drilling contractor's day rate is the major daily cost in the drilling phase. Most drilling contractors have considerable resources and this can be exploited by building in equipment into the drilling contractor specification which would otherwise require third party service payments. Items which can be successfully built into such contracts include:

non-magnetic drill collars
single shot survey equipment
crossovers
stabilisers
drilling jars
fishing tools
shaker screen consumption
cement tubing strings

For land work it is rarely economical to use a dedicated casing crew. The mobilisation cost and daily cost can never be recovered by the additional speed and efficiency, if any, that these crews achieve when the overall rig day rate is considered. Consequently, it should be clearly stated in these drilling contracts, that casing will be run by the drilling contractor alone.

The contractors should be asked to provide fuel consumption figures for their rig as this is traditionally an area where the operator foots the bill. Fuel consumption is another reason why it is wrong to go for a 'bigger is better' policy when selecting a rig, also a smaller rig is cheaper to mobilise.

No expenses should be paid to the contractor by the OPCO for the following and the drilling contract will make this clear:

rope, soap and dope consumption
all slip and elevator sizes required by OPCO
all tong jaw sizes required by the OPCO
all rubber goods
all wellhead test equipment

The contract should be constructed simply on the following basis:

Operating rate for all normal operations
Repair rate for first six hours of repair per well
Zero rate for subsequent repairs and contractor negligence

20. Procure materials

Material selection and purchase is standard on most land wells and does not, in most cases, require the expense of a materials man. Purchases required cover:

wellhead
bits (sale or return)
mud chemicals (sale or return)
casing

21. Select third party services

mud logging
electric logging
coring
directional drilling services, if required
transportation

22. Build site

It normally takes some four weeks to tender and award the site civil works contract. Thereafter, it may take six to eight weeks for site construction, depending on requirements.

23. Mobilise rig

In Northern Europe, mobilisation will normally be a matter of days, but may be dependent on the rig finishing work for another company. Site utilities should be available shortly before the rig is due to arrive.

24. Well operations

The choice of Superintendent and Supervisor is critical to the safe and speedy completion of the drilling phase. For a short well, assign one Supervisor to the site to cover the entire well. The Supervisor should be fully conversant with all aspects of the well design and prospects. Other factors to note are as follows:

Electric logging should be kept to an absolute minimum as the costs are proportionally higher than the rest of the evaluation process.
Mud engineer Should only be used if the mud system is too complex for the Drilling Supervisor to handle. Otherwise do not use a Mud Engineer.
Bits: use consignment bits as far as possible.
Coring: use consignment equipment as far as possible. Investigate the availability of part worn coreheads as coreheads represent the larger expenditure in the coring programme.
Directional policy: maintain vertical hole by good drilling practices. Run minimum surveys to confirm hole angle and direction. Run multishots at casing setting depths only.
Communications should require only three phone lines to the rig site:

one for the company rep. office
one for contractor TP
one for fax

The fax machine is cheaper and more effective than a telex machine. Anyone can operate it and all necessary reports can be transmitted quickly and clearly.

25. Test/suspend/complete/abandon

This is dependent on findings.

26. Restore site

all planning requirements must be met
any hedges or trees removed should be replanted
the site may require up to three years attention to satisfy local authorities.

Chapter 14

OFFSHORE DRILLING PROJECT MANAGEMENT

Managing offshore drilling projects is broadly similar to managing onshore projects. However, there are usually more third party services required, more paperwork required from the authorities, and furthermore it will probably cost a lot more and can take a lot longer. The secret, of course, is the planning and preparation. If this is thought out well ahead of time, then the operation should run smoothly. Most wells will require a three-month programming time, but HP/HT wells could take six months.

In this section we will look at a flow chart for the project and then discuss each part of it individually. UK North Sea legislative requirements have been used and if your particular well in your particular area does not require a specific item on the flow chart then it should be deleted from the outset. Conversely, if there is additional legislation or more requirements then these must be inserted into the flow chart.

1. PLANNING OFFSHORE DRILLING

Pre-planning

1. Selection of drilling prospect
2. Getting partner approval
3. Construction of planning bar chart

Detailed planning

4. Write programme, AFE and manuals
5. Request permission from authorities
6. Determine rig, service company and material requirements
7. Order long lead time items
8. Establish supply base
9. Perform site seabed survey
10. Receive approval from authorities

Drilling well

11. Tender for rig
12. Tender for services
13. Tender evaluation
14. Rig selection
15. Third party services selection
16. Mobilise rig
17. Position rig
18. Drill hole
19. Request permission to test from authorities
20. Well test
21. Request permission to clear site
22. P & A/complete/suspend
23. Perform seabed survey
24. Move rig off

	ACTIVITY	TIME IN MONTHS													
	PRE-PLANNING	1	2	3	4	5	6	7	8	9	10	11	12	13	14
1	SELECTION OF DRILLING PROSPECT	◄													
2	GET PARTNER APPROVAL	X◄													
3	CONSTRUCT PLANNING BAR CHART	X													
	DETAILED PLANNING														
4	WRITE PROGRAM A.F.E & MANUALS	X	XX	XX	X	◄									
5	REQUEST PERMISSION FROM AUTHORITIES		XX	XX	X										
6	DETERMINE RIG/SERVICE/EQUIP/NEEDS		X	XX	X										
7	ORDER LONG LEAD TIME ITEMS			X	XX	XX	XX	XX	XX	XX	XX	◄			
8	ESTABLISH BASE			X	XX	XX	XX	XX	XX	XX	XX	◄			
9	PERFORM SITE / SEABED SURVEY			XX	XX										
10	RECEIVE APPROVAL FROM AUTHORITIES				X	◄									
	DRILLING WELL														
11	TENDER FOR RIG					XX	XX	◄							
12	TENDER FOR SERVICES					XX	XX	◄							
13	TENDER EVALUATION							XX	◄						
14	RIG SELECTION							XX	XX	◄					
15	THIRD PARTY SERVICES SELECTION							XX	XX	◄					
16	MOBILISE RIG									XX	X	◄			
17	POSITION RIG										X				
18	DRILL HOLE											XX	XX		
19	REQUEST PERMISSION TO TEST												X	◄	
20	TEST													X	
21	REQUEST PERMISSION TO CLEAR SITE													X	◄
22	P & A / COMPLETE/ SUSPEND														X
23	PERFORM SEABED SURVEY														X
24	MOVE RIG OFF														X

Figure 1 Norwell flow chart for offshore drilling

1.1 Pre-planning

1. Raise prospect
This is done internally by the Explorationists.

2. Get partner approval
This is done by the company management.

3. Construct planning bar chart
At this stage, the best available estimate of the time it will take to achieve each function of the well plan should be laid out in a flow chart.

1.2 Detailed planning

4. Write manuals, AFE, Drilling Programme
In order to gain permission from the authorities to drill, a detailed programme and a full set of manuals will have to be submitted. The AFE is also required at this time to cost out the well and to indicate when the cash calls will be. (See chapters 5 and 7).

(a) The following manuals will need to be prepared (see chapter 4):

Emergency Contingency Manual (see chapter 4)
Operations Manual (based on Policy Document (see chapter 3))
Safety Manual (based on Policy Document (see chapter 3))
UKOOA Procedures* (buy in)
DOE Rules and Regs* (buy in)

The two manuals marked with an asterisk can be bought in from companies who specialise in providing just this type of manual. It is difficult for a small operator to keep abreast of UKOOA and DOE rules and regulations without assigning a lot of manpower to the task. It is normally a lot more cost-effective just to buy in a current manual and then keep it constantly updated during the operational period. Even a lapse of six months between drilling programmes could imply a lot of legislative changes and it is **vital** that the operator is working with a totally up-to-date set of guidelines, rules and regulations at all times.

(b) AFE should be constructed as shown in chapter 5.

(c) Drilling Programme should be written as shown in chapter 7.

5. Request permission from authorities
Within the UKCS, the following permissions are required:

drilling approval
establishment of 500 m safety zone around well location
approval to move a rig
approval to use oil-based mud

6. Determine rig service company and materials requirements

At this stage, we know sufficient about the well and the drilling location to be able to establish rig, service company and materials requirements. Rig selection has been covered in chapter 9 and we apply these basics laid down there to select a rig type, class or model. The service company requirements for an offshore well are manifold and the following list gives an indication of probable services required:

- site survey
- supply base
- supply boats
- helicopters
- offshore medical support
- weather forecasting
- standby boats
- communications
- rig positioning
- rig towing
- drilling consultants
- mud
- cementing
- mud logging
- directional drilling
- coring
- logging
- casing running
- geologists
- divers

7. Order long lead time items

If the well is to be drilled into high pressure reservoirs or into reservoirs which present us with additional problems such as H_2S or CO_2, then very special casing types and wellhead types may be required. If this is the case, then it must be anticipated that such equipment has to be specially manufactured or sourced. This, of course, takes time.

8. Establish supply base

An operational base must be set up as near as possible to the offshore well location. In UK waters, this will usually mean Lowestoft, Great Yarmouth or Aberdeen. These are the established ports which can provide the range of services to support an operation. By placing the operational base in one of these ports, the size of base required can be minimised. This is because these places are so organised that equipment can be left at the suppliers own depots until shipping, rather than collecting it in the operator's yard and warehouse. Storage space obviously costs money and by organising logistics the actual space leased by the operator can be kept to a minimum.

9. Perform site seabed survey

At this time, a boat will be sent out to the location to check the seabed conditions. What it is looking for mainly is to check for obstructions, determine seabed make-up, strength and flatness. From this information, the type of rig which can safely operate in the area can be confirmed. Furthermore, any special anchoring requirements or modifications can be planned for and accommodated.

10. Receive approval from authorities

Until this approval has been received, any well expenditure should be done with caution. In most cases in most fields, approval will be automatic, however, should there have been a recent incident, or if you plan to perform non-standard operations, then delays or rejection could be anticipated.

1.3 Drilling well

11. Tender for rig

Having determined the rig requirement and received official approval for the well, it is now time to go to the market to select a suitable rig.

12. Tender for services

As for the rig, third party services now need to be tied down contractually.

13. Tender evaluation

At this stage, the bids received from the contractors are assessed financially and technically to establish the most suitable choice to meet the well's objectives safely, efficiently and economically.

14. Rig selection

In the event that there is no obvious choice of rig, it will be necessary to inspect the 'front runners' and use the results of this inspection to influence the final decision. Even if there is a clear leader, the rig should be inspected by OPCO staff to ensure that it is what it purports to be and that has the equipment that it claims to have.

15. Third party services selection

At this point, the selection is made and contracts are issued for signing.

16. Mobilise rig

Ideally, the rig should already be in the area of interest. Whether it is or not, the organisation of the tow to the site will normally come down to the operator. The time taken to mobilise a rig will depend on its readiness for operation and its design.

Drillships will sail to a location themselves, however, all other rig types will be towed by a tug. Even self propelled semisubmersibles use a towing bridle with an 8000+HP tug to take them to the new location. When towing close to shore, or when moving through shipping lanes, it is a legislative requirement for semis to be under tow and not in self-propelled mode.

17. Position rig

Rig positioning is done using a combination of geostatic satellites and shore-based beacons. In the middle of the North Sea, the rig can be placed with an accuracy of around one meter radius. This should be accurate enough to satisfy most geologists.

18. Drill hole

At this time, the well is spudded and drilled according to the Drilling Programme.

19. Request permission to test from authorities

If the well drilled looks promising and worthy of testing, permission must be granted by the authorities prior to commencing testing. On offshore wells, testing is usually done in cased hole, so there is always a delay between finding a zone of interest and physically being in a position to test it. Draft programmes for testing should be ready along with safe working procedures and testing manual. If this is the case, then approval can be obtained very quickly.

20. Well test

The well is then tested to evaluate its following capabilities.

21. Request permission to clear site

From the evaluation of test data, a decision will be made internally by the operator and their partners as to the commercial viability of the well. Based upon this decision, the well will be completed, suspended or abandoned. If any equipment (xmas tree or wellhead) is to be left standing proud of the seabed, prior permission from the authorities must be gained prior to the rig leaving the location.

22. P & A/complete/suspend

Depending on well findings, one of the above operations will be carried out.

23. Perform seabed survey

An ROV is launched which checks the seabed area immediately below the rig for junk, debris or potential obstacles. Prior to leaving the location, all of the above must be removed and the seabed restored to its original condition.

24. Move rig off

Once the seabed is seen to be clear, the rig can be moved to the next location.

INDEX

Note: Numbers in *italics* indicate figures